Lecture Notes in Electrical Engineering

Volume 1063

The book series *Lecture Notes in Electrical Engineering* (LNEE) publishes the latest developments in Electrical Engineering—quickly, informally and in high quality. While original research reported in proceedings and monographs has traditionally formed the core of LNEE, we also encourage authors to submit books devoted to supporting student education and professional training in the various fields and applications areas of electrical engineering. The series cover classical and emerging topics concerning:

- Communication Engineering, Information Theory and Networks
- Electronics Engineering and Microelectronics
- Signal, Image and Speech Processing
- Wireless and Mobile Communication
- Circuits and Systems
- Energy Systems, Power Electronics and Electrical Machines
- Electro-optical Engineering
- Instrumentation Engineering
- Avionics Engineering
- Control Systems
- Internet-of-Things and Cybersecurity
- Biomedical Devices, MEMS and NEMS

For general information about this book series, comments or suggestions, please contact leontina.dicecco@springer.com.

To submit a proposal or request further information, please contact the Publishing Editor in your country:

China

Jasmine Dou, Editor (jasmine.dou@springer.com)

India, Japan, Rest of Asia

Swati Meherishi, Editorial Director (Swati.Meherishi@springer.com)

Southeast Asia, Australia, New Zealand

Ramesh Nath Premnath, Editor (ramesh.premnath@springernature.com)

USA, Canada

Michael Luby, Senior Editor (michael.luby@springer.com)

All other Countries

Leontina Di Cecco, Senior Editor (leontina.dicecco@springer.com)

**** This series is indexed by EI Compendex and Scopus databases. ****

Sanjay Yadav · Harish Kumar ·
Pavan Kumar Kankar · Wanyang Dai ·
Fenghua Huang
Editors

Proceedings of 2nd International Conference on Artificial Intelligence, Robotics, and Communication

ICAIRC 2022

Springer

Editors
Sanjay Yadav
MSI, CSIR-NPL
CSIR-NPL
New Delhi, India

Harish Kumar
Department of Mechanical Engineering
National Institute of Technology Delhi
Delhi, India

Pavan Kumar Kankar
Department of Mechanical Engineering
Indian Institute of Technology Indore
Indore, Madhya Pradesh, India

Wanyang Dai
Nanjing University
Nanjing, China

Fenghua Huang
Fuzhou Economic Technology
Development Zone
Yango University
Fuzhou, China

ISSN 1876-1100 ISSN 1876-1119 (electronic)
Lecture Notes in Electrical Engineering
ISBN 978-981-99-4553-5 ISBN 978-981-99-4554-2 (eBook)
https://doi.org/10.1007/978-981-99-4554-2

This Springer imprint is published by the registered company Springer Nature Singapore Pte Ltd.
The registered company address is: 152 Beach Road, #21-01/04 Gateway East, Singapore 189721, Singapore

Paper in this product is recyclable.

Contents

About the Editors

Prof. Dr. Sanjay Yadav born in 1962, obtained his master's degree in science (M.Sc.) in 1985 and his Ph.D. degree in Physics in 1990. Presently, he is working as the Editor-in-Chief (EIC) of the MAPAN: The Journal of Metrology Society of India. He is also Vice President of the Metrology Society of India (MS), New Delhi as well as Vice President of the Ultrasonic Society of India (USI), New Delhi. He is the Former Chief Scientist and Head of the Physico Mechanical Metrology Division of NPL and also a Former Professor at the Faculty of Physical Sciences, Academy of Scientific and Innovative Research (AcSIR), HRDG, Ghaziabad. He had taught the 'Advanced Measurement Techniques & Metrology' course, taken practical classes, and supervised graduate, master, and Ph.D. students since 2011. He is the recipient of research scholarships from the Ministry of Home Affairs, India (1986); CSIR, India (1988); Col. G.N. Bajpayee Award of Institution of Engineers, India (1989); Commendation Certificates from Haryana Government (1991 & 1992); JICA Fellowship of JAPAN (1998), Commendation Certificates from SASO, Saudi Arabia (2003); 3 Appreciation Certificates from Director, NPL (2005); Managing Editor, MAPAN (2006-2014); nominated as Member of APMP Technical Committee of Mass Related Quantities (TCM), Australia (2013–2019); Nominated as Country Representative in APMP, China (2019); Vice President, Metrology Society of India (2020); Member, National Advisory Committee, NCERT, Delhi (2019); Members, Testing and Calibration Advisory Committee, BIS (2019, 2020 and 2021), and very recently received a prestigious International award i.e. APMP Award for Developing Economies, China (2020). He has significantly contributed to pressure metrology, biomedical instrumentation, ultrasonic transducers and instrumentation systems. His current research interests include research and developmental activities in physicomechanical measurements; establishment, realization, maintenance and up-gradation of national pressure and vacuum standards; dissemination of national practical pressure scale to users through apex level calibration, training and consultancy services; interlaboratory comparisons, proficiency testing program and key comparisons, implementation of Quality System in the laboratory as per ISO/IEC 17025 standard and Finite Element Analysis (FEA) and Monte Carlo Simulations for pressure balances.

He has published more than 450 research papers in national and international journals of repute and conferences, 20 books, 14 patents and copyrights, supervised 8 Ph.D.s (another 5 in waiting), and drafted several projects, scientific and technical reports, documents and policy papers.

Harish Kumar is currently working as Associate Professor at the National Institute of Technology (NIT) Delhi. With more than 17 years of research and academic experience, he is an active researcher in the area of mechanical measurement and metrology. He served as a Scientist at the National Physical Laboratory, India, for more than a decade, and as a guest researcher at the National Institute of Standards and Technology, USA, in 2016. He has authored more than 90 publications in peer-reviewed journals and conference proceedings. He is an active reviewer for many prominent journals on measurement, metrology, and related areas and is an Associate Editor of the *MAPAN/Journal of Metrology Society of India*, published by Springer.

Pavan Kumar Kankar is the Associate Professor in the Department of Mechanical Engineering at the Indian Institute of Technology (IIT) Indore. His research interests include fault diagnosis and prognosis using machine learning techniques, reliability, bio-medical signal processing, and analysis of dynamic systems. He has over 20 years of teaching and research experience. Dr. Kankar obtained his Ph.D. from IIT Roorkee. He has published over 150 papers in refereed journals and conferences, with more than 3740 citations. He is an Editorial Board member of The *International Journal of Life Cycle Reliability and Safety Engineering.* He had served as a guest editor in the special issue of various journals. He is a member of professional bodies like the American Society of Mechanical Engineers, the Society for Reliability and Safety, the Tribology Society of India, and the International Institute of Acoustics and Vibration.

Wanyang Dai is the Professor at the Department of Mathematics of Nanjing University (doctoral supervisor/important subject post), Chief Scientist of the Dip's Institute of Digital Economy, and Chairman of Jiangsu Probability and Statistics Society. He has made several important series in the fields of quantum cloud computing, stochastic network reflection-diffusion approximation, stochastic (asymptotic) optimal control and (stochastic differential) game theory, and (forward-backward and reflection) stochastic (normal/partial) differential equations. He gave the keynote speech of the special guest conference at the international academic conference, was invited to make a special report at the 7th China Probability and Statistics Annual Conference in 2002, and was invited to make a special report at the 10th China Operations Research Annual Conference in 2016 which invented the network and computer cloud platform operating system and database that support various decision-making and big data analysis, and realized the separation of the network platform, intelligent engine, and end-user to achieve plug-and-play (Plug-Play) in an all-round way.

Fenghua Huang (Prof./Ph.D./Postdoctor) is the vice dean of the College of Artificial Intelligence, Yango University (China), director of the UERCSDMA (University Engineering Research Center for Spatial Data Mining and Application) in Fujian Province (China), director of the Institute of spatial data mining in Yango University, a master supervisor in the fields of computer technology, surveying and mapping engineering in Fuzhou University (China). He was a visiting scholar at the University of North Carolina (UNC) in the USA in 2017 and 2018 and was awarded as the excellent teacher of Fujian Province in China (2015–2017) in 2017. He is a member of IEEE, CCF, and CIE. In 2015, Dr. Huang was selected for the cultivation program for outstanding young scientific research talents at Fujian University. In 2016, he has selected into the new century excellent talents support program of Fujian universities and the overseas high-end visiting scholar program for outstanding discipline leaders of Fujian undergraduate universities. His research interests include machine learning, spatial data mining, big data, and remote sensing image processing. In the last decade, he has presided over more than 10 research projects funded by governments and enterprises. He has published more than 30 papers in related journals and conferences, including 15 papers indexed by SCI/EI and 4 papers indexed by CSCD. In addition, he has obtained the authorization of 8 national utility model patents and published 3 monographs and textbooks.

The Study for the Influencing Factors of the Box Office and Prediction Based on Machine Learning Models

Shengyi Chen, Shuyu Ni, Zhihan Zhang, and Zibo Zhang

Abstract Without a derivative market, film investments primarily rely on box office receipts as their primary source of revenue. For this reason, early box office forecasting is crucial because it enables investors to anticipate the project's overall success in advance and raise awareness of the movie using the right publicity techniques to boost its overall box office take. In this study, we first extracted a dataset containing 217 missing values that were suitable for analysis with regression and classification models. Then the data are preprocessed, including missing value processing, normalization, weakly correlated data processing and feature scatter analysis. If there is missing data, the data will be deleted or replaced based on whether the missing variables in the observed values are missing. Random forest model, neural network model, decision tree model, gradient lifting tree and extreme random forest are used to express the error measure between observations of the same phenomenon with the average absolute error. In addition, the evaluation index called MAE was used to evaluate the error based on the machine learning models. Finally, through the analysis of the experimental results and the mean absolute error size, it is found that the random forest model has the best performance and the smallest error in predicting the movie box office.

These authors contributed equally.

S. Chen
Suzhou No.10 High School of Jiangsu Province, Suzhou 215000, China

S. Ni
Yancheng No.1 High School Cambridge International Department, Yancheng 224000, China

Z. Zhang (✉)
School of International Education, Xuchang University, Xuchang 461000, China
e-mail: 15311020115@stu.cpu.edu.cn

Z. Zhang
Ealing International School, Shenyang 110013, China

S. Yadav et al. (eds.), *Proceedings of 2nd International Conference on Artificial Intelligence, Robotics, and Communication*, Lecture Notes in Electrical Engineering 1063, https://doi.org/10.1007/978-981-99-4554-2_1

1 Introduction

When a movie is launched, the box office refers to the sales revenue from the sale of movie tickets. The box office revenue of cinemas accounts for 80% or more of the overall revenue of cinemas in the current operational stage of Chinese cinemas. Therefore, it is very important for the feasibility analysis of cinema investment and cinema management, including cinema trusteeship to be able to predict the cinema box office income. This paper will analyze the commonly used formula of box office prediction and its constituent elements. Every year thousands of films are shown around the world, and the box office income of a film is an important indicator of the success of a film. High box office films can not only bring huge profits, but also reflect the high standards of film directors, actors and producers. However, the audience's taste in the film is unpredictable, leading to huge risks in the investment of film companies in new films. Therefore, box office prediction is a research direction that the film industry pays more attention to, especially whether accurate box office prediction can be given as soon as possible. If the box office can be predicted as early as possible, the film producers and distributors can make appropriate adjustments to the production and distribution of the film according to the box office prediction data, and use the budget more reasonably to obtain higher profits. As an important indicator to measure the profitability of a film, the box office is affected by many factors and its influencing mechanism is complex, so it is difficult to predict the box office accurately. This project uses an open-source film dataset to build a box office prediction model. Firstly, factors affecting the box office, such as film type, release schedule, director and actor, are quantified and analyzed visually. The box office of a film is affected by a variety of factors.

Some studies have been done to do the box office prediction. For instance, Subramaniyaswamy et al. employed the linear regression and SVM method for predicting the box office success (Subramaniyaswamy 2017). In addition, Liu et al. predicted movie box office revenues based on the linear and non-linear regression models (Liu 2016). The experimental results demonstrated the effectiveness of the methods. However, the performance of these methods still has a certain space to be further improved. In addition, the comparison of various machine learning methods is limited.

To solve the limitation mentioned above, this paper studies the influencing factors of the box office and divides the influencing factors into three categories. In addition, serval typical machine learning models were employed to train the collected dataset. it is found that the random forest model has the best performance and the smallest error in predicting the movie box office.

2 Method

2.1 Dataset Description and Processing

The dataset used in this paper is from archive.ics.uci.edu and CSM (Traditional and Social Media Movies) dataset in 2014 and 2015 (UCI machine learning repository 2015 CSM ... 2015). The attributes of the dataset include features e.g. Movie, Year, Ratings and Genre. The dataset has 14 multiple features, including 217 data, with missing values, which is suitable for regression and classification model analysis.

In general, there are a lot of incomplete (with missing values), inconsistent and abnormal data in the original data. Therefore, data preprocessing is particularly important. By showing the number of missing values in each column of attributes in the data set, it is found that there is at least one missing value in each attribute, which is mainly concentrated in the Screens and Aggregate Followers. By directly deleting the missing label data and filling the missing feature data with mode, we cleaned the data and completed the deletion and filling of each attribute data.

Normalization is a way of simplifying computation, which is to transform a dimensionless expression into a dimensionless expression, a scalar. In this paper, the Min–max method is used to normalize data, which is also called deviation standardization. This method is a linear transformation of the original data, so that the resulting values are mapped to between (0 and 1). The conversion function can be found in for

$$x' = \frac{x - \min(x)}{\max(x) - \min(x)} \tag{1}$$

The degree to which two variables are related is referred to as correlation. In general, the scatter plot shows that there is either a positive correlation, a negative correlation, or an uncorrelation between two variables. In this paper, through the correlation analysis of each feature attribute and label attribute, the heat map of correlation coefficient is obtained as shown in Fig. 1.

It is obvious from the figure that the two attributes with the least correlation with Gross are Year and Sentiment. Delete the two columns; Also, Movie is not directly related to the purpose of our box office forecast and can be deleted.

2.2 Proposed Approach

This paper compares the Random forest model, Decision tree model, Extra trees model, Gradient boosting model and Neural network model.

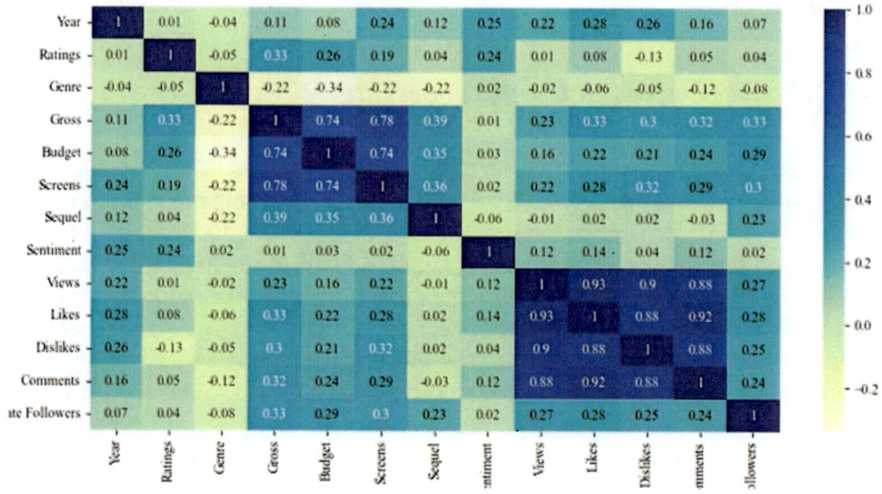

Fig. 1 The correlation coefficient

2.2.1 Random Forest Model

In this paper, Random forest model is used to predict the data. Random forest is an integration method designed for decision tree classifiers, which is an extension of bagging method. Random forest and bagging method adopt the same sampling method. Bagging method of decision tree each time from the selection of all attributes in the attribute as a branch of an optimal attribute, and Random forest algorithm every time a randomly selected from all the properties F attributes, and then from the F in selecting an optimal attribute as its branch attribute, which makes the entire model of randomness stronger, making stronger generalization ability of the model. The selection of parameter F determines the randomness of the model. If there are M sample attributes in total, $F = 1$ means that one attribute is randomly selected as the branch attribute. When F is equal to the total number of attributes, it becomes the bagging method integration. The weak classification decision tree used by random forest algorithm is usually CART algorithm. Random forest algorithm is simple and easy to implement, and it has a good classification effect.

The advantages of Random forest are presented as follows:

- Missing values are not taken into account. The accuracy can still be preserved even if a significant portion of the details are lost;
- During training, trees are independent from each other, and the training speed is fast;
- The randomness is introduced and it is not easy to overfit.

The parameters used in the study of random forest are max depth is set to 30, min samples split and min samples leaf are set to 2 and random state is set to 60. In addition, the mean absolute error is selected as the evaluation index to evaluate the performance of the model.

2.2.2 Decision Tree Model

Decision trees are an important machine learning method (Myles 2004). Algorithms for creating decision trees include ID3, C4.5, and CART. In a decision tree, each internal node corresponds to an attribute judgment, each branch to the output of the judgment, and each leaf node to the classification outcome.

2.2.3 Extra Trees Model

Since each decision tree uses the original training set and is highly aggressive, Extra Trees often do not use random sampling (Sharaff et al. 2019). Instead, they will randomly select a feature value to divide the decision tree.

2.2.4 Gradient Boosting Decision Tree Model

Gradient Boosting Decision Tree (GBDT) is an iterative Decision Tree algorithm, which is mainly used for regression. After improvement, GBDT can also be used for classification tasks. GBDT builds multiple decision trees and synthesizes the output results of all decision trees to get the final result (Cheng et al. 2018).

2.2.5 Neural Network Model

In a BP network, additional layers (one or more layers) of neurons are added between the input layer and the output layer. These hidden units have no direct external connections, but changes in their state can have an impact on how the input and output are related. Each layer may contain multiple nodes (Wang 2003; Vohradsky 2001). The superiority of the neural network has been fully demonstrated in many researches (Yu 2022; Zhou 2022). In this study, the batch size and epochs are set to 200 and 150, respectively.

3 Result and Discussion

3.1 Result

From the comparison in Figs. 2, 3 and 4, it can be seen that the predicted value and the real value of random forest fit better, especially at some peak values. Compared with other models, random forest has higher line coincidence and lower mean absolute error. However, the common problem of several models is that the fit is not strong when x is between 5 and 15 and after 50.

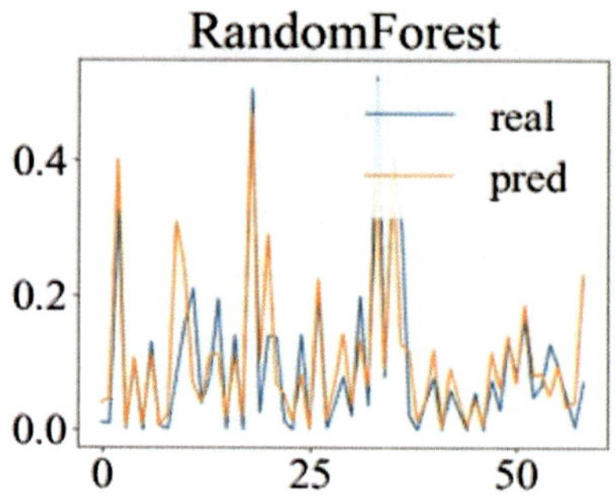

Fig. 2 Comparison of performance of Random forest

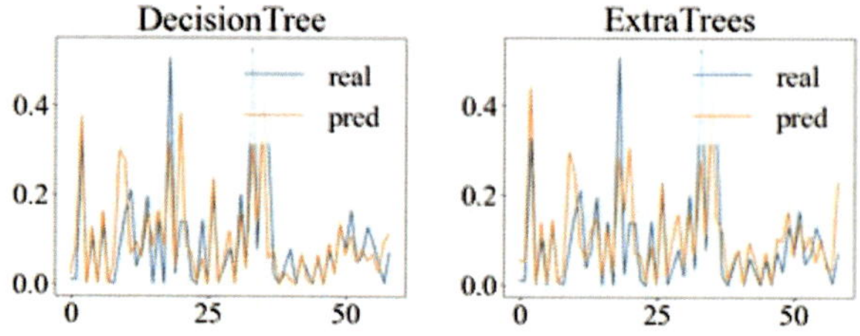

Fig. 3 Comparison of performance of Decision tree (left) and extra trees (right)

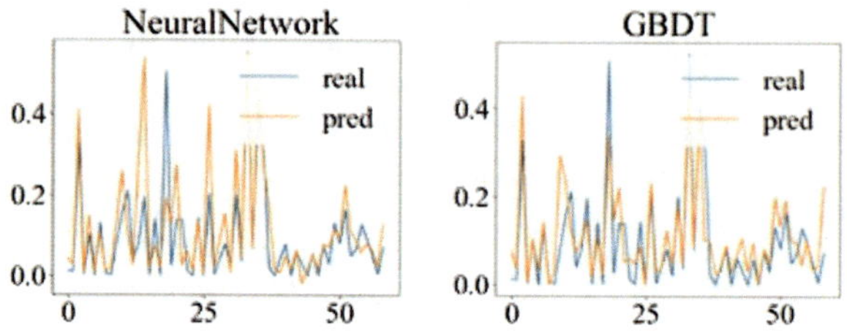

Fig. 4 Comparison of performance of GBDT (left) and neural network (right)

3.2 Discussion

The reason why the prediction result of random forest is better is that it carried out the comprehensive decision due to the structure of multiple decision trees compared to other models. Additionally, during the tree-building process, when nodes are segmented, the best segmentation point is chosen among a random selection of characteristics rather than the best segmentation point among all genera. Because of this randomness, a random forest will typically have a higher bias than a single non-random tree, but because it is averaged, its variance will also be lower, typically offsetting the bias increase and producing a better model overall. We adopt the control variable method when determining the model parameters. Adjust each parameter of the model and observe the performance of the model, select the parameters with the best performance of the model, in order to achieve a more accurate prediction effect.

4 Conclusion

In this study, we compared the neural network and several regression tree models, and finally determined that the random forest model had the best performance and the smallest error value in predicting the movie box office. At the present stage of the study, only the data set of traditional and social media movies in 2014 and 2015 is used to compare the model predictive value with the real value. Later, it is intended to obtain more recent movies or movies that have not been shown for prediction, and gradually develop into a prediction system. In addition, mining based on the above-mentioned big data related to films will provide decision-making reference for roles in different links of the film industry. In the future, a more advanced model e.g. Transformer will be considered in further study to improve the performance based on the current version.

References

Cheng J et al (2018) Research on travel time prediction model of freeway based on gradient boosting decision tree. IEEE Access 7:7466–7480

Liu T (2016) Predicting movie box-office revenues by exploiting large-scale social media content. Multimed Tools Appl 75(3):1509–1528

Myles AJ (2004) An introduction to decision tree modeling. J Chemom: J Chemom Soc 18(6):275–285

Sharaff A et al (2019) Extra-tree classifier with metaheuristics approach for email classification. In: Advances in computer communication and computational sciences. Springer Singapore, pp 189–197

Subramaniyaswamy V (2017) Predicting movie box office success using multiple regression and SVM. In: 2017 international conference on intelligent sustainable systems (ICISS). IEEE, pp 182–186

UCI machine learning repository 2015 CSM (Conventional and Social Media Movies) dataset 2014 and 2015 data set. https://archive.ics.uci.edu/ml/datasets/CSM+%28Conventional+and+Social+Media+Movies%29+Dataset+2014+and+2015

Vohradsky J (2001) Neural network model of gene expression. FASEB J 15(3):846–854

Wang SC (2003) Artificial neural network. In: Interdisciplinary computing in java programming. Springer Boston MA, pp 81–100

Yu Q (2022) Pose-guided matching based on deep learning for assessing quality of action on rehabilitation training. Biomed Signal Process Control 72:103323

Zhou Q et al (2022) Multi-modal medical image fusion based on densely-connected high-resolution CNN and hybrid transformer. Neural Comput Appl, 1–21

Structured Model of "Three Flows in One" Emergency Preplan Based on Knowledge Graph

Hongbin Li, Chao Liu, Yong Li, and Huilan Zeng

Abstract Emergency preplan can be viewed in the guidance document to respond to emergencies. To improve its effectiveness and applicability, this dissertation raised the structural model and method of the "Three flows in one" convergence emergency preplan based on a knowledge graph, which, constructed on the "Three flows in one" of the structural decomposition and connection of scenario flow, task flow, element information flow, aims to achieve the digitalization and intelligentization of emergency preplans. In terms of improving the decision support ability of emergency preplans, promoting the management system and modernization of emergency, it is of great practical significance.

1 Introduction

In an emergency system with Chinese characteristics and with "one preplan and three regulations" as its core, the emergency preplan stands as the basis of emergency management, the most pivotal process to reach the digitalization of emergency preplans, and also the data basis to prompt the digitalization of emergency preplans. At present, most researches incline to put information technologies into a systematic platform to make an application, yet comparatively less researches are made on core contents of structuralization of emergency preplans, such as preplan process,

H. Li
Safety Supervision Department, State Grid Corporation of China, Jing County, Anhui, China

C. Liu
Safety Supervision Department, State Grid Corporation of China, Linyi County, Shanxi, China

Y. Li
Department of Electricity Industry Management Consultancy, Shanghai Jiulong Enterprise Management Consultancy Co. Ltd., Xuzhou, China

H. Zeng (✉)
Department of Electricity Industry Management Consultancy, Shanghai Jiulong Enterprise Management Consultancy Co. Ltd., Ji'an, China
e-mail: zenghuilan2008@163.com

S. Yadav et al. (eds.), *Proceedings of 2nd International Conference on Artificial Intelligence, Robotics, and Communication*, Lecture Notes in Electrical Engineering 1063, https://doi.org/10.1007/978-981-99-4554-2_2

model decomposition, etc. (Jingjing et al. 2018). To further optimize the guidance and controlling ability, it is of great necessities to promote the structuralization of emergency preplans and enhance the response ability to emergencies.

2 Analysis on Pain Points of Emergency Preplans

Proactive explorations have been made on the structuralization and digitalization of emergency preplans. Despite certain achievements acquired, the whole practice still displayed relative weakness. Most explorations tended to decompose the organization's responsibility and classical emergency resources, but they lacked accuracy and utility in the response process and emergency tasks (Yingshun et al. 2022). The major problems go as follows:

A. Lack of disposal schemes, which leads to difficulty to execute preplan in some way.
B. Lack of utility. Some preplans in the past indeed made some requirements in disposal methods, yet they did not connect those methods with required teams, materials, equipment and other resources.
C. Deficient in digitalization of preplan. The past preplans, with trivial processes, did not form the structural ones or target to designate a person precisely, which led to the slow response to preplan.

3 Definition and Advantages of Structuralization of Emergency Preplans

The structuralization of emergency preplans means that to compose the text contents of emergency preplans according to a specific way to form some kinds of structural elements that can be recognized by computers as well as the combination based on certain relations and orders, which shall apply digitalization technologies into preplan implementation. As for the structuralization of preplan, it is the core process of emergency preplans to transform the text into digitalization.

The structuralization of emergency preplans endows remarkable advantages, it is easy to conduct scenario analysis, detailed in disposal methods and with strong maneuverability and utility.

4 Structural Models of "Three Flows in One" Emergency Preplans Based on Knowledge Graph

The structuralization of "Three flows in one" emergency preplans based on knowledge graph is demonstrated in the graph below. The knowledge system of emergency preplans would be built based on knowledge graph first, and then the structuralization of emergency preplans with task process as the main line would be conducted for the structural decomposition and connection of three flows (structural decomposition first, and then structural connection), namely, scenario flow, task flow and element information flow. There are three important processes in general in the structuralization of emergency preplans: structural decomposition of scenario, structural decomposition of proposal task and creation of task flow, and connection of element information flow (see Fig. 1).

4.1 Knowledge System of Emergency Preplans Constructed Based on Knowledge Graph

Knowledge graph is a technical method to describe knowledge and build relations among diverse things in the world by graph. From its essence, it can be seen as the calculable model of relations among things, with the basic purpose to build the direct relations of knowledge for swift search and decision-making.

(1) Construction of knowledge graph of emergency preplans implemented jointly by two levels (guidance level and implementation level).

The emergency preplan implemented jointly by two levels means that, under specific disaster, the guidance level and implementation level would put the preplan into practice accordingly. For guidance level, it will issue different task orders or disposal requirements for executors with different identities under various scenarios according to the task process; and for implementation level, to complete the task order or disposal requirements handed by the guidance level, it will choose or change appropriate task implementation packages every now and then for different scenarios that are triggered until the task will reach the completion standard issued by guidance level or receive the task termination order from the guidance level.

The knowledge graph constructed in this dissertation can satisfy the needs of emergency preplans implemented jointly by two levels. This graph supports the top-down construction method and is also a construction method from down to top with interconnection between guidance level and implementation level.

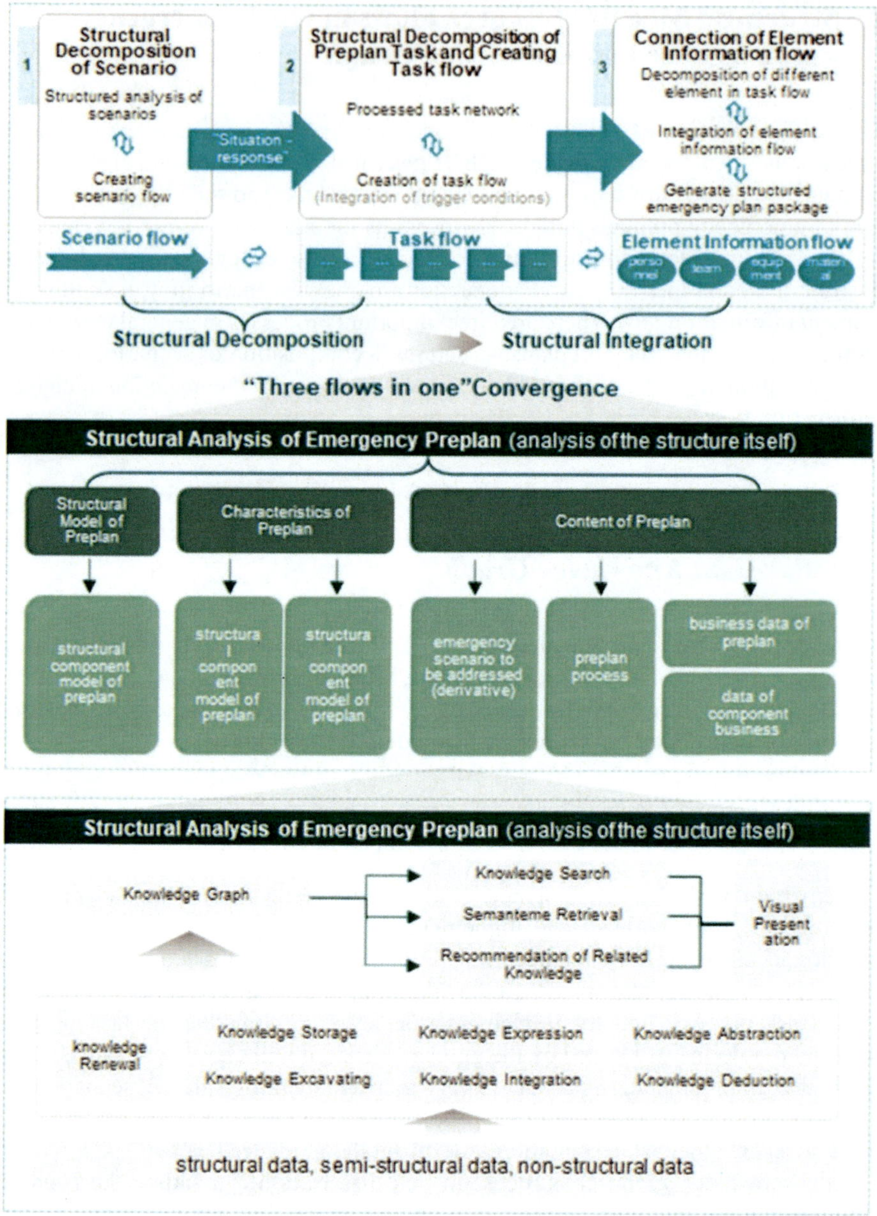

Fig. 1 Structural model of "Three flows in one" emergency response proposal based on knowledge graph

(2) Knowledge system of emergency preplans constructed based on knowledge graph.

The knowledge system of emergency preplans is a process that can be updated ceaselessly. During renewal, the knowledge system can form the knowledge graph through knowledge storage, knowledge expression, knowledge abstraction, knowledge excavating, knowledge integration knowledge deduction and other processes, so as to serve for knowledge search, semanteme retrieval, recommendation of related knowledge and visual presentation, etc. (see Fig. 2).

4.2 Disassembly of Emergency Preplan Structure

The structural disassembly of contents and features is required for the structural emergency preplans. The emergency preplan consists of the basic situation of preplan, responsibility of organization, scenario, task process of disposal based on preplan, resources and other structural elements.

4.3 Creating the Task Flow

The emergency task is usually witnessed as the combination of several sub-tasks.

The task flow is the major route to promote emergency preplan. Different task processes would be seen in different phases (such as the routine, early warning and emergency phases). In the structuralization of emergency preplans, great emphasis should be put on the task implementation of emergency preplans interconnected by two levels (guidance level and implementation level). And task flow should be executed by scenario.

(1) Multi-tiered process decomposition of preplan task: based on the combination between principal and flexibility and between systematical view and logical view, the multi-tiered process decomposition of preplan task should be conducted (one-tired, two-tiered or three-tiered decomposition). And regulations should also be made to create the multi-kind tasks and abstraction of task relations (Wenyan 2020), while combing through the preplan task processes, specifying each key point and related personnel and detailed disposal methods that should be made simultaneously, so as to form the task flow with specific divisions of responsibilities.
(2) Creating and promoting preplan task flow.

Combined with practical disposal experience and fully combing through emergency scenarios at different phases such as early warning and response, the preplan should be decomposed into a combination of different scenarios, so as to connect guidance

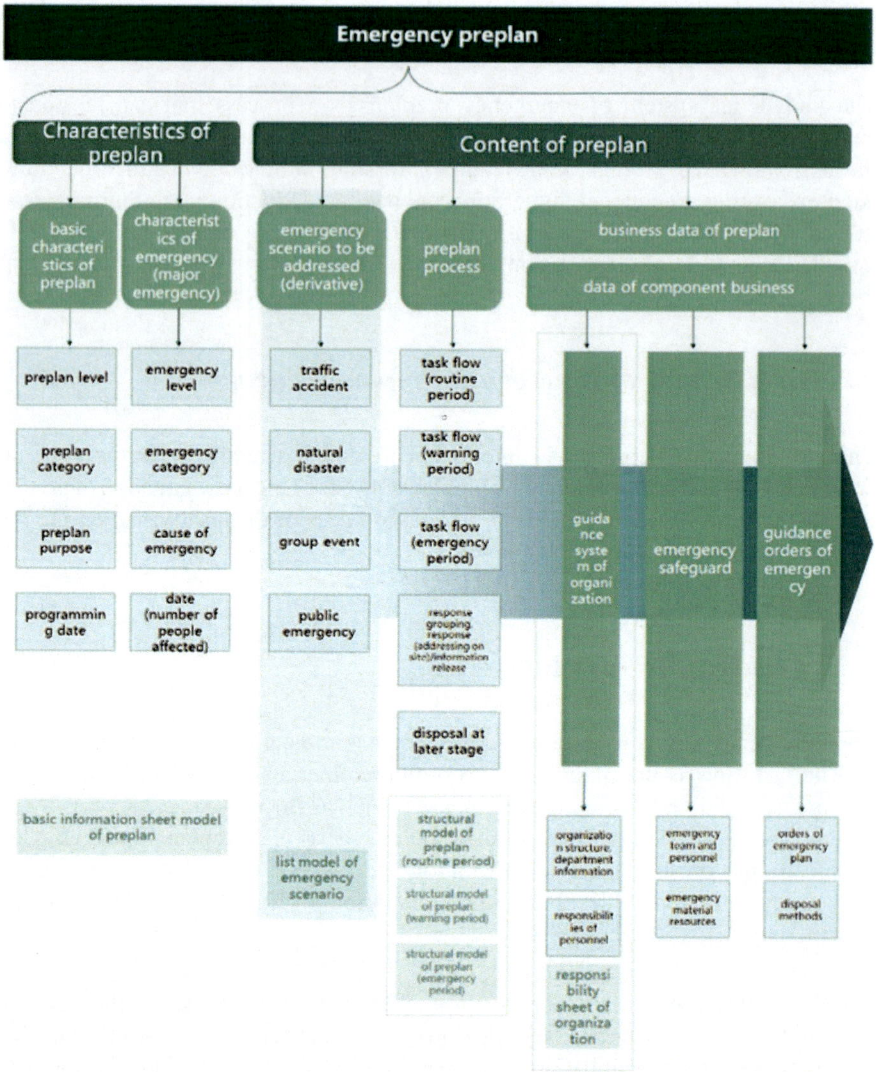

Fig. 2 Knowledge system of emergency preplans constructed based on the hierarchical structure of knowledge graph

and task methods. Efforts should be made to create a task flow (connected with numerous elements) that goes through the whole process and the whole procedure. The trigger should be specified, and the logic of task solution module and emergency works should also be formed, so as to closely connect the organs and connect the task procedure with guidance elements. By this way, the linkage among each scenario and automatic switch among different procedures can be ensured, which is conducive to making emergency guidance disposal task packages and connect.

(3) Workflow Engine Technology Application and Flow Chart Construction of Emergency Plan Tasks.

For the task process of the emergency plan, the BPMN (Business Process Model and Notation) business process model and symbol specification conforming to the workflow engine Activities are used to model the flow chart in the process of disassembling the task flow, so as to use the flow chart disassembling tool to structurally disassemble it later.

For each active node on the task flow, it is also necessary to use an attribute table (XML technology) for assignment definition, endow each node with code, and structurally associate with the corresponding scene element, trigger condition, instruction element, action element and resource element.

4.4 Task Flow Driven by Scenario

Structural analysis of scenario: given the complexity and rationality, structural analysis of scenario should be made on different kinds of scenarios that can be involved with preplan, and scenarios should be used to embody the relations among various issues in unconventional emergencies (Minghong 2016), so as to make a scenario database of emergency that can support multi-scenario combinations in structural preplan for better utility.

Creating scenario flow and responding: to construct scenario flow connected with key points in a task flow. The preplan should be driven by scenario, under which disposal orders shall be issued by guidance personnel accordingly, to cope with different emergency scenarios flexibly.

4.5 Connection of Element Information Flow

The information resources of emergency response are comprised of city basic data, danger origin, emergency safeguard resources, disaster data of the site, etc., which is the information collection created by each involved entity and related to emergency response work (Chen et al. 2022). The element information can be decomposed structurally into personnel, team, material resource, equipment, etc.

The multi-element information flow connection should be made based on the main line of task flow to form the structural preplan package similar to a script, so as to help people with different responsibilities to know "what to do, "how to do" and "what they shall use".

5 Application of "Scenario-Response" Assisted Decision-Making of Structured Emergency Plan

The functional architecture of the "scenario-response" auxiliary decision application module of structured emergency plan is shown in the figure below.

The system combines the structured emergency plan of "three in one" and the command measures, decision-making rules and other contents of emergency command and disposal, based on the emergency deduction operation engine, according to the preset resources and the current situation, it can give multiple disposal action plans recommended by the system, and then it can select the best emergency command and disposal plan recommended by the system by sorting the plans in combination with the emergency disposal goals (see Fig. 3).

With the situational evolution of emergency situation flow and the implementation of emergency command and disposal task flow, the "scenario-response" auxiliary

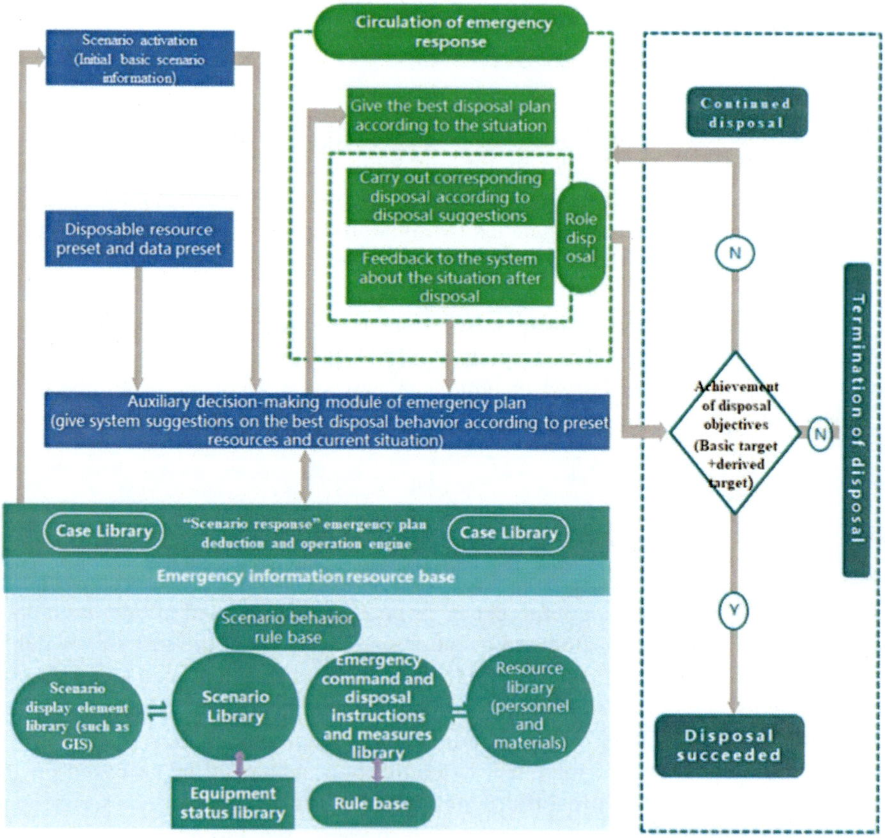

Fig. 3 Application of "scenario-response" auxiliary decision-making of structured emergency plan

decision application module of a structured emergency response plan is equivalent to a circulation body of emergency command and disposal, which will continue to provide the best plan until the emergency command and disposal goal is finally achieved.

6 Case Experiment

By constructing the knowledge graph, the structuralization of emergency preplans implemented jointly by two levels can be consolidated. To verify the feasibility of the "Three flows in one" structuralization of emergency preplans based on knowledge graph, two preplans were chosen for experimental analysis, namely, an emergency preplan for the fire disaster in a substation and another emergency preplan for flood and typhoon prevention. Combined with the complexity of these emergency preplans, the corresponding knowledge graphs were built automatically, namely, the emergency disposal knowledge graph of the fire disaster in the substation, and three emergency disposal knowledge graphs of flood and typhoon prevention (which differs in the routine period, warning period and emergency period).

During the experimental analysis, the text of emergency preplan was worked as the input, and the structuralization of "Three flows in one" emergency preplan was conducted combined with the knowledge graph interconnected with two levels. Through all these processes, in the structuralization of emergency preplans, the reduction rate of preplan programming time was about 35%, the increase rate of knowledge database was around 40%, and the directness of the navigation diagram of preplan operation increased by 50%, which also indicated the remarkable enhancement of the utility and timeliness of the structuralization of "Three flows in one" emergency preplans. The scheme generation speed based on the "scenario-response" auxiliary decision module was increased by 80%.

7 Conclusion

The structuralization of preplan is the key of digitalization and intelligentization of emergency preplans. Implemented conveniently with high effectiveness, the structural model and methods of "Three flows in one" structural emergency preplan based on knowledge graph is of great practical significance to enhance the decision support ability of emergency preplans.

References

Chen Y, Bo F, Yupan Z (2022) Research on constraining factors and model innovation of multi-entity emergency response information synergy. J Mod Inf 42(7):31–41

Jingjing W, Wei Z, Yukun G (2018) Research on the structuralization of emergency preplans and flow lining technological scheme under the view of digitalization. J Saf Sci Technol 14(12):164–168

Minghong Z (2016) Deduction method research on scenario of unconventional emergencies based on examples. Huazhong University of Science and Technology, Wuhan, p 52

Wenyan G (2020) Quality assessment of emergency preplans response text abstracted based on process model and research on its revision method. Shandong University of Science and Technology, Qingdao, p 32

Yingshun L, Jing R, Jingjing W (2022) Discussion on development orientation and approach of digitalization of emergency preplans. China Emerg Manag 9:20–31

A Novel Action Recognition Method Based on Attention Enhancement and Relative Entropy

Xing-Sheng Liu and Shi-Jian Liu

Abstract Action recognition is a hotspot field in computer vision for video understanding. To achieve accurate and real-time action recognition, a novel skeleton-based method combined with attention enhancement and relative entropy is proposed. Firstly, the data normalization method is presented to reduce the recognition error caused by human body posture and scale differences. Secondly, a joint attention strategy is proposed considering the different contributions of joints in an action. Finally, a relative entropy based acceleration strategy is employed to avoid unnecessary matching and save time. Experiments carried out on the Microsoft Research Cambridge-12 dataset show that the accuracy of our method is 90.32% and the time cost is 31.58 ms on average, which outperforms the others.

1 Introduction

As a hotspot field in computer vision, human action recognition is widely used in video surveillance (Elharrouss et al. 2021), medical rehabilitation (Yan et al. 2020), etc. To achieve that, one way is to use textural features extracted from video frames (Qiu et al. 2017), while others take skeleton sequences as input for recognition (Cao et al. 2017). Figure 1 illustrated a skeleton sequence captured by MS Kinect. Due to the high-level abstraction of the human body, skeleton-based methods are superior to textures-based methods in terms of robustness and computational load in general. Therefore, this paper focuses on action recognition using skeleton sequences.

There are majorly three types of skeleton-based action recognition methods, which use (1) graph convolution, (2) statistics and (3) template matching respectively. For instance, a spatial–temporal graph convolutional network is proposed by Zheng et al. (2019). One downside of the graph convolution based method is that it requires a large number of data to train the model well. Besides, high computation and storage

X.-S. Liu · S.-J. Liu (✉)
Fujian Provincial Key Laboratory of Big Data Mining and Applications, Fujian University of Technology, Fuzhou 350118, Fujian, China
e-mail: liusj2003@fjut.edu.cn

S. Yadav et al. (eds.), *Proceedings of 2nd International Conference on Artificial Intelligence, Robotics, and Communication*, Lecture Notes in Electrical Engineering 1063, https://doi.org/10.1007/978-981-99-4554-2_3

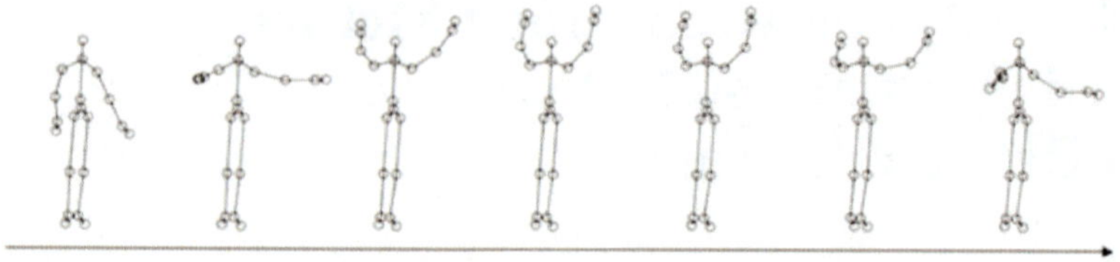

Fig. 1 Illustration of the skeleton sequence data

requirements make it hard to scale on resource-limited remote devices. Mostly used statistical methods include the Hidden Markov Model (HMM) (Choi and Kim 2018) and the Support Vector Machine (SVM) (Nguyen and Le 2015). The core strategy of template matching is to measure the similarity between the input and the given templates. And output the action type of the template that is most similar to the input. The popularly used similarity metrics are the results from the dynamic time warping (DTW) algorithm and its variances. This work chooses the DTW-based template-matching method because it is more flexible than others. Major contributions can be summarized as follows.

- We showed that normalization of the skeleton data is beneficial for system performance.
- A joint attention mechanism is presented to cooperate with the improved LDTW algorithm proposed in our previous work.
- To accelerate the recognition speed, a relative entropy based strategy is proposed.

2 Preliminary and Problem Statement

As shown in Fig. 2, the skeleton used in this work consists of 20 key points (i.e., v_1–v_{20}) and 19 vectors (i.e., e_1–e_{19}). Given a sequence of skeleton data, our work aims to find a meaningful action that it represented.

3 Proposed Method

As shown in Fig. 3, our method could be divided into 4 steps, which are data normalization with invariant features, extraction of joint attention, template filtering based on relative entropy, and matching respectively.

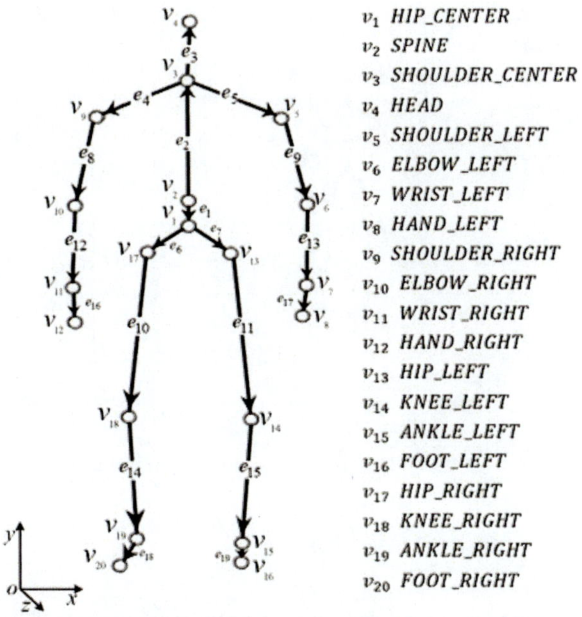

Fig. 2 The definition of the skeleton

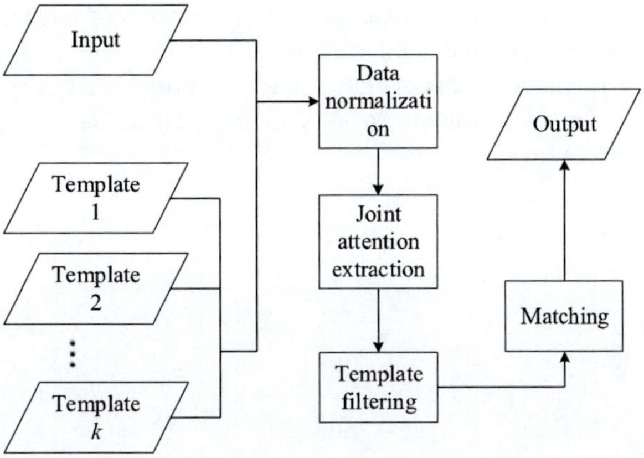

Fig. 3 The workflow of the proposed method

3.1 Data Normalization

Data normalization is a key step to improve accuracy because a skeleton may be different from the others in terms of position and scale in the 3D space. To avoid the affection of these discrepancies, we normalized the skeletons as shown on the left of Fig. 4 to the uniformed ones as depicted on the right. The normalization includes moving the skeletons to a uniform position (see the star in Fig. 4) and scaling the limbs to standard lengths.

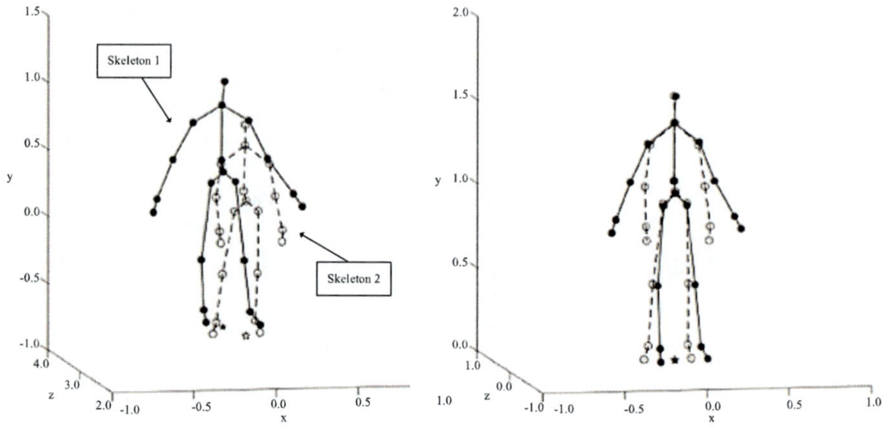

Fig. 4 Two skeletons before (left) and after (right) data normalization

3.2 *Joint Attention*

When matching the input with templates using the DTW algorithm, the contributions of each joint in the similarity evaluation are identical traditionally. However, taking the waving as an example, the trajectory of each joint in Fig. 5 shows that the elbow, wrist and hand joints are more important than others to distinguish this action. Motivated by this insight, a joint attention strategy is proposed to weigh each joint.

Given a sequence of N skeletons, let $\overline{\alpha}_k$ be the normalized weight of v_k, which meets Eq. 1.

Fig. 5 The trajectory of each joint for waving

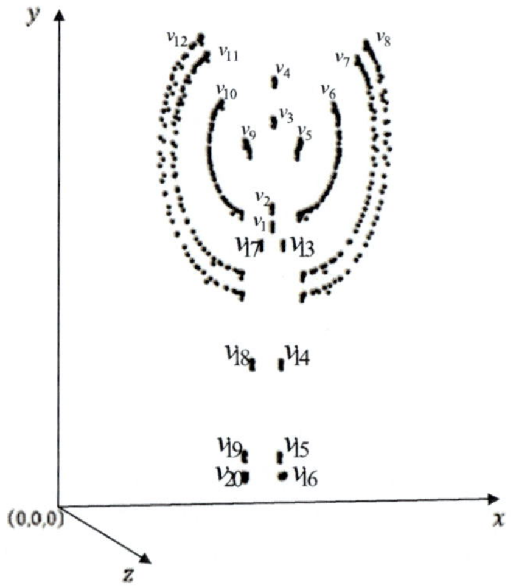

Fig. 6 The distribution of the proposed joint weights of 4 actions

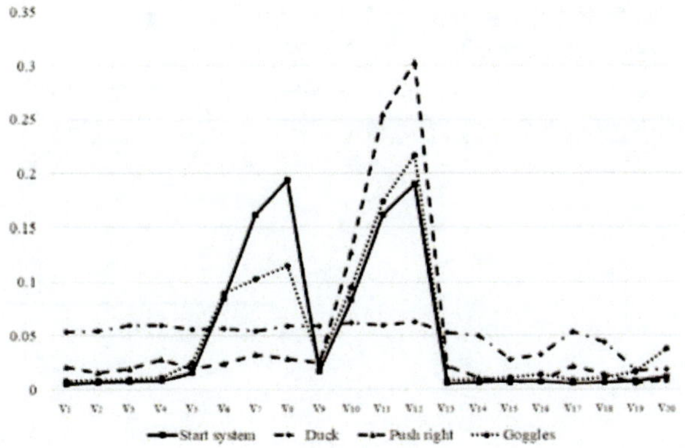

$$\begin{cases} \alpha_k = \sum_{i=1}^{N-d} \left\| v_{i+d,k} - v_{i,k} \right\|_2 \\ \overline{\alpha}_k = \frac{\alpha_k}{\sum_{j=1}^{a} \alpha_j} \end{cases} \quad (1)$$

where d is set to 10 empirically, a is the total number of joints. Figure 6 showed the distribution of the proposed joint weights of actions named "Start system", "Duck", "Push right", and "Goggles" respectively. By counting on the specific joints with higher weight, the system can tell an action from the others easily.

3.3 Acceleration Strategy

An improved LDTW algorithm (Zou et al. 2022), previously proposed by us, is adopted to fulfill the similarity calculation. To avoid involving it between the input and each template, an acceleration strategy is proposed to filter out some templates based on the relative entropy. Given the input X, and a template Y, if the relative entropy $D(X||Y)$ between X and Y as shown in Eq. 2 is larger than a threshold λ, Y won't survive to participate in the next DTW-based matching.

$$D(X||Y) = \sum_{k=1}^{a} \overline{\alpha}_k(X) \times \log_2 \frac{\overline{\alpha}_k(X)}{\overline{\alpha}_k(Y)} \quad (2)$$

where $\overline{\alpha}_k(X)$ and $\overline{\alpha}_k(Y)$ is the proposed weight of v_k in X and Y respectively. Figure 7 demonstrated the proposed filtering results of 12 typical actions by setting $\lambda = 0.2$.

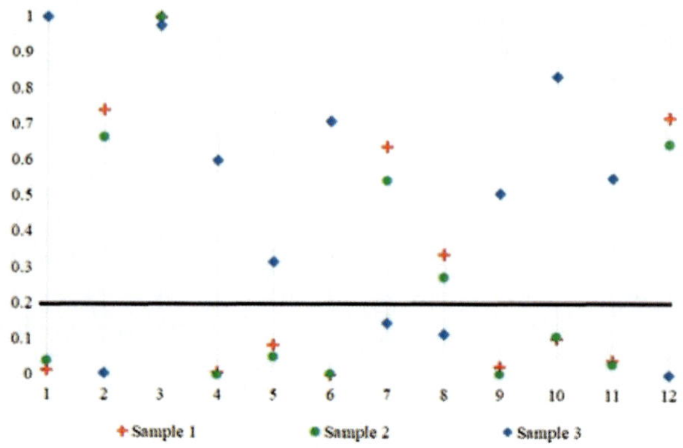

Fig. 7 Relative entropy based filtering results

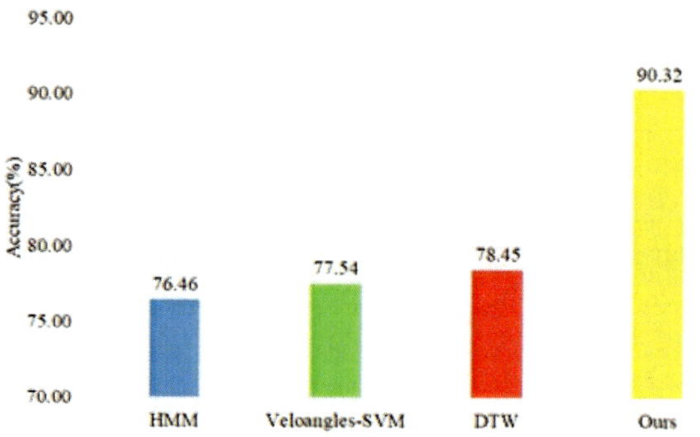

Fig. 8 Comparison results in terms of accuracy

4 Experiment and Results

All experiments were carried out on a desktop with AMD Ryzen 7 5800X CPU, and 64 GB memory with C++ as the programming language. The dataset we used is the MSRC-12 (Fothergill et al. 2012) which consisted of 596 sequences in 12 actions.

The first experiment is to compare our method with HMM (Choi and Kim 2018), Veloangles-SVM (Nguyen and Le 2015) and DTW (Sempena et al. 2011) in terms of accuracy, which is measured as the correctly recognized samples over the total samples. The results are shown in Fig. 8, where our method is the best compared to the others.

To validate the efficiency of the proposed acceleration strategy, we record the time cost with $\lambda = 0.1$, $\lambda = 0.2$ and without the template filtering respectively. Figure 9 shows with the help of the proposed acceleration, our method could achieve real-time recognition (i.e., frame rate $\geq$30 fps) when $\lambda \leq 0.2$.

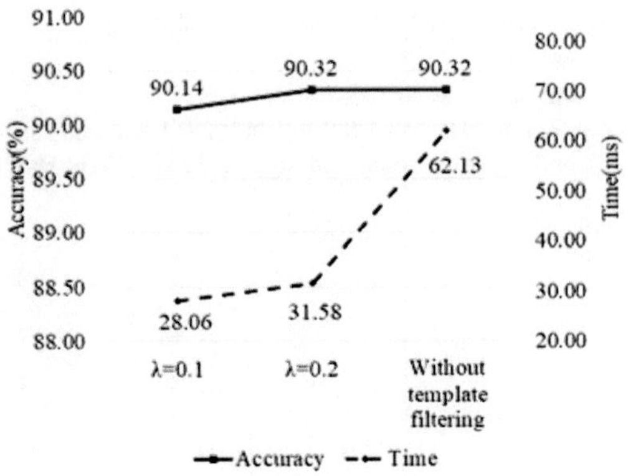

Fig. 9 Evaluation in terms of speed

5 Conclusion

This paper presents a novel method for action recognition using skeleton sequences. The core idea is to consider the contribution of specific joints when fulfilling the action recognition with DTW-based template matching. Besides, data normalization is involved to improve accuracy and relative entropy is adopted for acceleration. The experiments showed the effect and efficiency of the proposed method.

References

Cao Z, Simon T, Wei SE, Sheikh Y (2017) Realtime multi-person 2d pose estimation using part affinity fields. In: Proceedings of the international conference on computer vision and pattern recognition, pp 7291–7299

Choi HR, Kim T (2018) Modified dynamic time warping based on direction similarity for fast gesture recognition Math. Probl Eng 2018

Elharrouss O, Almaadeed N, Al-Maadeed S, Bouridane A, Beghdadi A (2021) A combined multiple action recognition and summarization for surveillance video sequences. Appl Intell 51(2):690–712

Fothergill S, Mentis H, Kohli P, Nowozin S (2012) Instructing people for training gestural interactive systems. In: Proceedings of the SIGCHI conference on human factors in computing systems, pp 1737–1746

Nguyen DD, Le HS (2015) Kinect gesture recognition: SVM vs. RVM. In: Proceedings of the international conference on knowledge and systems engineering, pp 395–400

Qiu Z, Yao T, Mei T (2017) Learning spatio-temporal representation with pseudo-3d residual networks. In: Proceedings of the international conference on computer vision, pp 5533–5541

Sempena S, Maulidevi NU, Aryan PR (2011) Human action recognition using dynamic time warping. In: Proceedings of the international conference on electrical engineering and informatics, pp 1–5

Yan H, Hu B, Chen G (2020) Real-time continuous human rehabilitation action recognition using OpenPose and FCN. In: Proceedings of the international conference on advanced electronic materials, computers and software engineering, pp 239–242

Zheng W, Jing P, Xu Q (2019) Action recognition based on spatial temporal graph convolutional networks. In: Proceedings of the international conference on computer science and application engineering, pp 1–5

Zou Z, Nie MX, Liu XS, Liu SJ (2022) Improved LDTW algorithm based on the alternating matrix and the evolutionary chain tree. Sensors 22(14):5305

Application of Combinatorics Based on Discrete Analysis in WCET Embedded Software Testing Technology

Zongling Yu, Bo Wang, Huachen Zhao, Chunxin Shi, and Zhe Xu

Abstract This paper focus on the analysis of the worst execution time of embedded software WCET and the research of software testing methods and execution. The embedded software runs on the microprocessor, collects and receives external instructions and status information data, perform logical operations, and drives external mechanical structures or components to work according to the specified actions. The execution time analysis is closely related to the actual running of the function call tree, microprocessor instruction cycle, executive action status conditions, etc. This paper studies the most accurate and most valuable worst execution time analysis method of WCET, it has been verified with the project.

1 Introduction

With the need for intelligent development, embedded software research and development has entered a new stage, and Software testing has also evolved from a very small number of unregulated jobs. Detailed and clear management specifications have been formulated, a professional software testing team has been formed, and clear software testing objectives have been formulated (Zhang and Xie 2018). For the research and development of embedded software, time performance is the foundation to ensure the normal operation of the program. Distinct from the function realization, function realization directly acts on the external actuator. The position of time performance analysis in software testing is comparable to that of software in the controller, which is invisible but plays a decisive role (Cao and Yin 2019). The benchmark of time performance analysis is the machine time used when the cycle executes the most code or the most complex code. Time performance analysis can

Z. Yu (✉) · B. Wang · H. Zhao · C. Shi
AVIC Xinxiang Aviation Industry (Group) CO. Ltd., Xinxiang 453000, China

Z. Xu (✉)
School of Mechanical and Electrical Engineering, Henan Institute of Science and Technology, Xinxiang 453003, China
e-mail: xuzhe@hist.edu.cn

S. Yadav et al. (eds.), *Proceedings of 2nd International Conference on Artificial Intelligence, Robotics, and Communication*, Lecture Notes in Electrical Engineering 1063, https://doi.org/10.1007/978-981-99-4554-2_4

not only analyze the impact of program running time on function implementation, but also optimize the software architecture (Ermedahl et al. 2011).

Worst Case Execution Time (WCET for short), as an important measure of time performance analysis, has established a certain research mode. At present, there are two kinds of outstanding methods in WCET research for embedded software: one is to develop corresponding mathematical models, simulate their operation, and find the longest path in combination with the types of microprocessor chips that are widely used in the market (Burns and Puschner 2000); On the other hand, the hardware device is used to grab the time mark of each program execution, calculate the running time of each program by calculating the time difference between the two time marks, and compare the maximum time. No matter which implementation method is selected, the program based WCET statistics will not change (Ha et al. 2016).

2 Analysis of the Tested Object

Combined with the tested object in this paper: embedded software code, that is, software program based on hardware microprocessor, has the following inherent attributes:

(a) The instruction cycle of the program must be provided by the time crystal oscillator of the microprocessor;
(b) The microprocessor must provide support for hardware registers, data space, program running space, etc.;
(c) Embedded software will be affected by external command signals, and will send commands to external actuators at the same time;
(d) The implementation of core functions of embedded software requires system initialization, self-inspection
(e) Support of software modules such as data acquisition;
(f) Embedded software has a power supply and power output;
(g) Embedded software includes digital quantity (low-voltage level signal), analog quantity input and output.

3 WCET Worst Execution Time Calculation Principle

The core of the worst execution time calculation of embedded software is based on the hardware platform. Grab the time signs of each node during each run, and find the longest and shortest time performance quantitative data according to the time signs (Zhang et al. 2002). The basic principle is clear, and the key point is how to run. In order to achieve the most correct longest and shortest time calculation, each branch and condition combination must be executed. On the premise of correct big data capture, the worst execution time of WCET can be obtained by comparing (Li et al.

2019). Use combinatorics to conduct multi-dimensional combination, at the same time, each subset of the combination is composed of judgment conditions. Possible branches of each judgment condition are Yes or No, which are regarded as discrete quantity (Zhao 2021).

Each condition is regarded as a set of discrete quantity results, and the discrete quantity is arranged arbitrarily. Assume that there is *A*, *B*, *C*, … *N* execution conditions, and each condition combines 2, 3, 4, 5, … , *m* conditions (Lim Hyun Bae 1995). The non-repeated combination type obtained by conditioning combination is:

$$A = N * (N - 1) * (N - 2) * (N - 3) * (N - 4) * \ldots * (N - (N - 1)) \quad (1)$$

The big data are obtained after the discrete quantities are arbitrarily arranged and combined. After the big data are executed, the time sign is obtained. The execution time of each time is calculated, from which the longest and shortest execution times are taken.

After all combinations are executed, the maximum time calculation formula is obtained:

$$\mathrm{WCET}_{\max} = \max_{1 \le m \le N} \begin{bmatrix} (T_{1\text{-first}} - T_{1\text{-end}}), \\ (T_{1\text{-first}} - T_{1\text{-end}}) \ldots (T_{A\text{-first}} - T_{A\text{-end}}) \end{bmatrix} \quad (2)$$

Minimum time calculation formula:

$$\mathrm{WCET}_{\min} = \min_{1 \le m \le N} \begin{bmatrix} (T_{1\text{-first}} - T_{1\text{-end}}), \\ (T_{1\text{-first}} - T_{1\text{-end}}) \ldots (T_{A\text{-first}} - T_{A\text{-end}}) \end{bmatrix}. \quad (3)$$

4 WCET Worst Execution Time Software Test Execution Standard

To achieve 100% code coverage, suppose that each action is executed, it needs to get through: data collection, data processing, patrol inspection, logic judgment (normal and abnormal logic processing), and action execution (Chattopadhyay 2012). Suppose there are N actions, and each action consists of two states. After combination, all the combined contents are obtained. When executing the test, it is necessary to create all the possibilities of a normal combination, and determine the expected output results according to the logic (see Table 1).

Table 1 Combined status of software Test Execution()

1	Number of actions executed each time (n)	1	2	3	4	5	……	n
2	Number of execution action combinations (m)	C_N^1	C_N^2	C_N^3	C_N^4	C_N^5	……	C_N^n
3	Create conditions for each action	X	Y	Z	P	Q	……	L

When executing the test case, you need to create the conditions in line 3 of the table above to get the action executed. Combine the executive actions and eliminate the combinations that cannot be executed at the same time to obtain feasible combination conditions. According to the analysis of functional requirements, quality characteristic requirements, performance requirements, interface provision, etc., get the results of combined conditional execution, and compare the actual results with the analyzed results to see whether the results are correct (Knoop et al. 2011).

If the actual output result is inconsistent with the expectation, it can be analyzed from the following four aspects: first, software code errors; Second, test case errors; Third, the operation steps are wrong; Fourth, demand errors.

Analysis principle: First, assume that the software requirements are correct, and analyze whether the code implementation is correct; If the code implementation is consistent with the requirements, analyze whether the test case is correct; If the use case is correct, analyze whether the operation or equipment has problems; Finally, we will discuss whether the demand is correct and whether it meets the customer's needs.

5 Examples of Engineering Software Testing

According to the code modules required for software function realization described in the chapter "WCET Worst Execution Time Software Test Execution Standard", select one of the modules to receive communication, and analyze the WCET Worst Execution Time for this functional module. Function requirement description: Receive the communication data sent by the upper computer, and perform data analysis and exception verification according to the provision of the communication protocol. According to the software functional requirements, design the software unit and use this as the architecture of software coding, see Fig. 1.

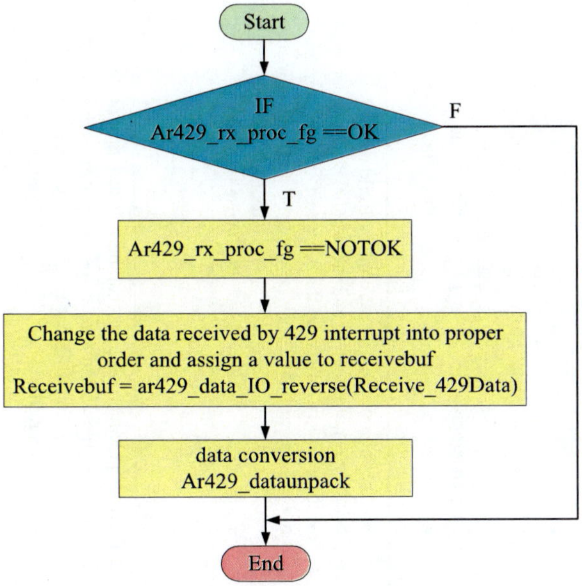

Fig. 1 Example of software design flow chart (owner-draw)

The worst execution time analysis of software WCET is as follows: First, build a test environment and use an oscilloscope to monitor the actual running time; Secondly, according to the software design flow chart, the instruction cycle value on the hardware processor is calculated theoretically; Third, use the WCET worst case execution time test tool to measure. Using the above three different methods, a worst case run time analysis was performed, as shown in Table 2.

Connect the test piece to the test fixture and monitoring computer to build the test environment, as shown in Fig. 2. According to the theoretical analysis, through the test tooling, the most complex test conditions are loaded with the test piece to ensure that the test piece runs in the worst execution state. In this state, the execution time of the functional module in the test piece is monitored by the oscilloscope, and the test results are shown in Fig. 3.

Theoretically, 430 lines of code are calculated, the instruction cycle is 6.67 ns, and 2.867 μs is calculated.

Combined with the function call tree, as shown in the figure, the tool measurement results are shown in Fig. 4.

Table 2 Summary of execution results of different software test methods (owner-draw)

Test method	Method description	WCET measurements (μs)	Operating steps	Influence factor	Confidence/bias	Key points of implementation
Method 1	Actual product testing	6.2	Build test environment; running products; perform the test, monitor the execution data and record; calculate the WCET value	Upper computer command transmission frequency; communication resource competition; oscilloscope accuracy	The execution of the highest/core procedure is not affected, and the deviation is only caused by the tools and operations	Build the difference analysis between the test environment and the real environment; tool calibration; tool use standard
Method 2	Numerical statistics based on design theory (about 430 lines)	2.867	Processor analyzes instruction cycle; code review; calculation of theoretical value	Consideration of other synchronous execution instructions in working state Instruction cycle	The lowest value/value that can be used as the most ideal state is too low, too ideal, and different from the actual operation	Instruction cycle confirmation during processor execution; Confirmation of consistency between software design and software code
Method 3	Use professional tools to measure	7.607	Determine the scope of analysis; conduct analysis configuration; entry function selection; confirmation of analysis results	Accuracy of tool model, tool pile insertion time	It is slightly better than Method 2, and the value is on the high side. Because the tool execution requires inserting pile code into the original program, the original code scale is increased, and the execution time is extended	The accuracy of the tool model The execution time occupied by the piling code

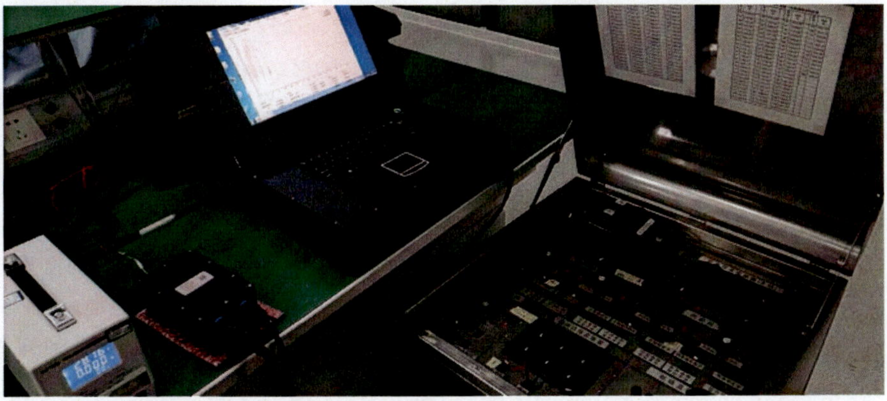

Fig. 2 Product test environment (owner-draw)

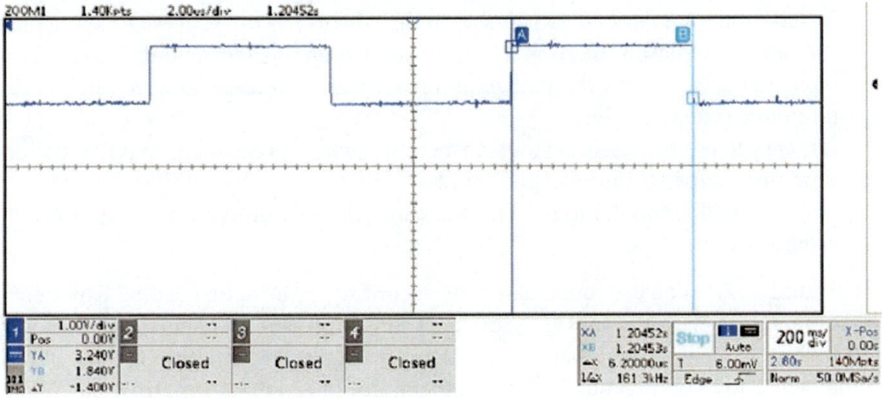

Fig. 3 Actual product test results (owner-draw)

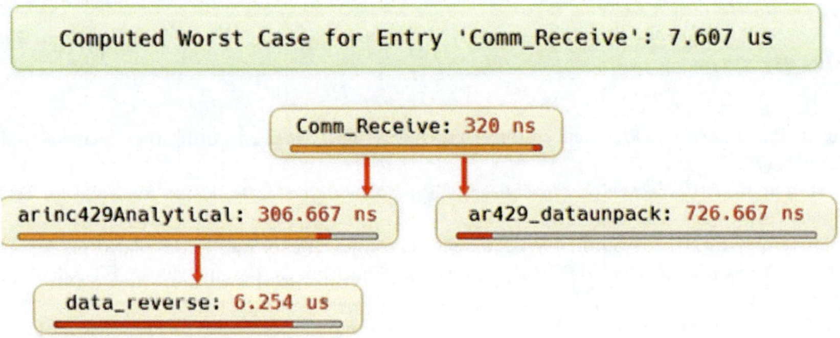

Fig. 4 measurement results using professional tools (owner-draw)

6 Conclusion

The worst execution time of WCET software testing uses combinatorics in overall planning. According to different tested objects, the combinatorics used can be discrete or continuous. The key point is "all", the branch executive should be complete, state identification should be complete, the combination analysis should be complete, and the demand scenario analysis should be complete. According to three different WCET worst execution time software testing methods, the analysis and summary are as follows:

1. In the actual product testing, the test results are the most accurate, but the external resource support, such as building a test environment, is the most difficult to meet in the implementation process of the project. In the actual testing process, the test is simple and difficult to test. To implement all the condition combinations, the preparation of test cases requires a lot of energy and time.
2. Based on the numerical statistics of design theory, in the process of static analysis and calculation, there is a lack of impact analysis of external operational interference data, and the calculation results are too ideal to have engineering reference research value.
3. It is easy to use professional tools to measure, which greatly shortens the working time and helps to improve the work efficiency. However, the test results are more dependent on the tools. The working principle of the tools determines the accuracy of the test.

To sum up, we can use tools to operate according to the actual testing experience, in combination with the tool measurement deviation and mechanism, and modify the tool measurement results in combination with the tool analysis results and the implementation coefficient. This method can not only greatly improve the WCET worst execution time software testing efficiency, but also ensure the testing accuracy.

References

Burns A, Puschner P (2000) Guest editorial a review of worst-case execution-time analysis. R Time Syst 18(2)

Cao M, Yin X (2019) Research on scenario based software test case design method. Sci Technol Innov Inf (14):74–75

Chattopadhyay S (2012) A unified WCET analysis framework for multi-core platforms

Ermedahl A, Bygde S, Lisper B (2011) An efficient algorithm for parametric WCET calculation. J Syst Arch 57(6)

Q Ha, D Liu, X Shen, Liu L (2016) Test case generation method for aerospace software based on requirement model. Opt Precis Eng 24(5):1185–1196

Li C, Jiang Y, Song Y, Feng L (2019) Design and application of airborne software testing tool chain. Comput Meas Control 27(6):55–61

Knoop J, Kovács L, Zwirchmayr J (2011) Symbolic loop bound computation for WCET analysis

Lim S-S, Hyun Bae Y (1995) An accurate worst case timing analysis for RISC processors. IEEE Trans Softw Eng 21(7)

Zhao Y (2021) Research on measurement model of military software testing process. North University of China
Zhang Y, Wang G, Yang P, et al (2002) Research on requirement based aviation embedded software testing technology. Comput Eng Des 23(10):4–7
Zhang Y, Xie D (2018) Research on embedded software testing technology based on model design. Electron World (9):66,68

Path Planning Based on Multi-parameter Adaptive Improved Elite Ant Colony Algorithm

Hu Dan and Huang Hui

Abstract The traditional elite ant colony algorithm has strong randomness, easy to fall into local optimal solution, slow convergence and other problems in path planning. This paper proposes that the information heuristic factor α expectation heuristic factor β, pheromone volatilization coefficient ρ, and the pheromone increase coefficient e of the optimal path is changed into a dynamic adaptive parameter, which changes with the number of iterations. In this paper, the grid method is used to model the path environment in space, and an adaptive parameter mathematical model is constructed and verified. By comparing the simulation experiments of different scale map environments, it is found that the improved elite ant colony algorithm has significantly improved the path optimization ability, smoothness and robustness.

1 Introduction

How to obtain excellent path planning ability for mobile robots has always been the focus of robot research. (Zhu and Mingzhong 2010) For this problem, scholars at home and abroad have done a lot of relevant research work, among which the traditional methods are artificial potential field algorithm (Wang et al. 2003), Dijkstra algorithm (Fadzli et al. 2015), etc. In addition, swarm intelligence algorithms such as particle swarm optimization algorithm (Wang et al. 2012), genetic algorithm (Wang et al. 2021a), ant colony algorithm (Ma and Qian 2021), and deep reinforcement learning (Wang et al. 2021b) have also been applied to path planning.

Ant colony algorithm simulates the behavior characteristics of ants searching for food in nature. The algorithm has good robustness and environmental applicability, but it has strong randomness, is easy to fall into local optimal solution, and has slow convergence speed. Therefore, many scholars have proposed relevant solutions to this problem. Reference (Liu and Liang 2021) improved the guidance factor based on the elite ant colony algorithm, and added the distance factor from the node to the starting

H. Dan (✉) · H. Hui
Taizhou Vocational Collage of Science and Technology, Taizhou 318020, Zhejiang, China
e-mail: hudandexinxiang@163.com

S. Yadav et al. (eds.), *Proceedings of 2nd International Conference on Artificial Intelligence, Robotics, and Communication*, Lecture Notes in Electrical Engineering 1063, https://doi.org/10.1007/978-981-99-4554-2_5

point to improve the disadvantage that the ant colony algorithm is prone to fall into local optimization, but the path obtained is not the optimal solution. Reference (Jing et al. 2019) constructs a new heuristic function and adaptive pheromone volatilization factor to enable the algorithm to converge quickly, but the final path obtained by the algorithm is not the best path. On the basis of reference (Xu et al. 2021), a multi-directional and multi neighborhood search method is proposed. Combined with the vector angle as heuristic information, the transition probability control parameter is introduced to improve the optimization and convergence of the algorithm. This paper proposes a new improved method based on the elite ant colony algorithm for mobile robot path planning in unknown environments.

2 Elite Ant Colony Algorithm

Ants in the ant colony algorithm will gather along the path with high pheromone concentration. If there are multiple unvisited grids in grid i in the map, ant k will select the location of the next grid j according to the transition probability formula of Eq. (1), α Represents a pheromone factor, β Represents a heuristic factor.

$$p_{ij}^{k}(t) = \begin{cases} \frac{\tau_{ij}^{\alpha}(t)\cdot\eta_{ij}^{\beta}(t)}{\sum\limits_{s\in \text{allowed}_k} \tau_{is}^{\alpha}(t)\cdot\eta_{is}^{\beta}(t)}, & j \in \text{allowed}_k \\ 0, & \text{otherwise} \end{cases} \tag{1}$$

Traditional ant colony algorithm has strong blindness and slow search speed at the initial stage of the algorithm. In order to improve the performance of the original ant colony algorithm, M Dorigo et al. put forward the elite ant colony algorithm. The elite ant refers to the ant whose path is the best among all ants from the beginning of the algorithm to the present, and accelerates the convergence of the algorithm by additional enhancement of the information on the optimal path selected by it. The algorithm calls the best path that has been found T^{bs} (best so far), and the corresponding pheromone concentration modification formula is:

$$\tau_{ij}(t+1) = (1-\rho)\tau_{ij}(t) + \Delta\tau_{ij}(t) + \Delta\tau_{ij}^{\text{bs}} \tag{2}$$

$$\Delta\tau_{ij}^{\text{bs}} = \begin{cases} \frac{e}{L^{\text{bs}}}, & \text{edge } (i, j) \text{ part of optimal solution found} \\ 0, & \text{otherwise} \end{cases}$$

$$e = Q \tag{4}$$

In formulas 1 and 2, L^{bs} is the length of the known optimal path T^{bs}; e is the optimal path pheromone increase coefficient; $\rho(0 < \rho < 1)$ indicates the volatilization coefficient of pheromone on the path; Q is the intensity coefficient of pheromone increase.

3 Establish Space Environment Model

This paper uses grid method to build robot environment model. The mobile robot uses its own depth camera to collect map data, collect information and make analysis, which can accurately determine which areas have obstacles.

As shown in Fig. 1, the grid map is numbered 1, 2, 3, … from left to right and from bottom to top. The grid number corresponds to the coordinates one by one. It is specified that the robot moves on the center point of the grid. The mathematical relationship between the coordinates and the grid number is shown in formula (5).

$$\begin{cases} x_i = \begin{cases} a\left[\mathrm{mod}(i, N_x) - \frac{a}{2}\right], & \mathrm{mod}(i, N_x) \neq 0 \\ N_x - \frac{a}{2}, & \mathrm{mod}\ (i, N_x) = 0 \end{cases} \\ y_i = a\left[\mathrm{ceil}\left(\frac{i}{N_y}\right) - \frac{a}{2}\right] \end{cases} \tag{5}$$

In formula 5, i represents the grid serial number, N_X and N_Y represent the number of rows and columns of the grid map, a represents the grid side length of a unit, mod() is the remainder operation, and ceil() is the upward rounding operation.

In order to prevent the robot from colliding with obstacles, when the irregular obstacles are less than one grid, fill them into a grid, and then expand the obstacles

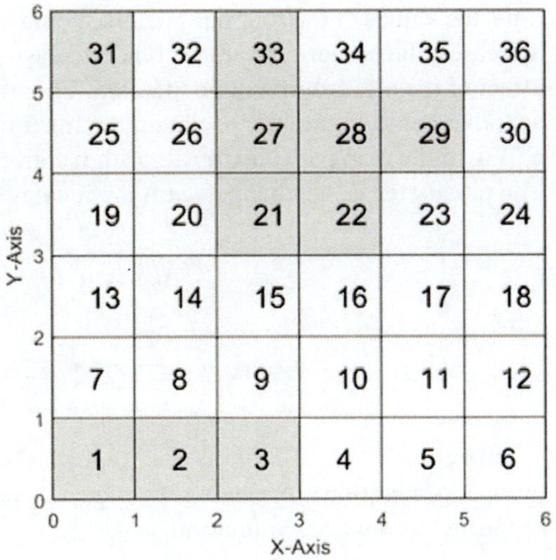

Fig. 1 Grid map

outward. The movable area is marked white in the grid diagram, and the value is 0, so the robot can pass; The black is the forbidden area, and the value is 1, indicating that there are obstacles on the path, and the robot needs to detour. A two-dimensional array matrix map (N_X, N_Y) is used to represent the obstacle map as follows:

$$\mathrm{map}(p, q) = \begin{cases} 1, & \text{obstacles in row} \, p \, \text{column} \, q \\ 0, & \text{otherwise} \end{cases} \tag{6}$$

4 Improved Elite Ant Colony Algorithm

Compared with the traditional elite ant colony algorithm, the parameters of the ant colony algorithm are improved: α (Information elicitation factor), β (expectation elicitation factor),ρ (Pheromone volatilization coefficient) and e (optimal path pheromone increase coefficient) are changed to dynamic adaptive parameters, which change with the number of iterations.

In traditional ant colony algorithm α and β is usually a fixed value in Zhu and Mingzhong (2010); Jing et al. 2019), ρ is $[0.1, 0.9]$, and the coefficient e is the value of the pheromone increase coefficient Q. But whether α, β, ρ or e, if they are too large, the pheromone influence weight on the path will be too large, making it easy for ants to enter the local optimal solution, thus losing the ability of global search; If the value is too small, the advantage of the best path cannot be reflected. The randomness of ant walking is too strong, and the convergence speed is too slow.

Therefore, at the beginning of the algorithm iteration, in order to avoid the algorithm falling into the local optimum and losing the global search ability of the algorithm, the pheromone of the optimal path should not be increased too much. At this time, the values of the four parameters should be smaller. As the number of iterations increases, the numerical value should increase gradually, otherwise, the convergence effect of the algorithm will be affected. Therefore, this paper selects the function of adaptive iteration number to obtain the maximum optimization of the global search ability of the map path and the algorithm convergence speed in the dynamic balance. The parameter value changes with the increase of iteration number k.

$$A(k) = \mu + \lambda \cdot \frac{k}{k + K} \tag{7}$$

$$e(k) = \lambda Q \cdot \left(\frac{k}{k + K}\right) + \mu Q \tag{8}$$

In formulas 7 and 8, K is the total number of iterations, and k is the current number of iterations. A means α or β or ρ. λ and μ are the value that can be adjusted according to the actual situation.

5 Experimental Simulation and Analysis

To verify the feasibility of improving the elite ant colony algorithm in this paper, software Matlab R2020b was used to simulate between traditional elite ant colony algorithm, algorithm in this paper, algorithm in reference (Liu and Liang 2021) and reference (Jing et al. 2019).

5.1 20 × 20 Grid Map Simulation Experiment

The 20 × 20 grid map in reference (Liu and Liang 2021) is used for the experiment of simulation data. The results are shown in Figs. 2 and 3, and the comparison of relevant evaluation index data is shown in Table 1. In the experiment, $Q = 40$.

It can be seen from Table 1 that the convergence path length of the traditional elite ant colony algorithm is 35.899cm, and there are 13 inflection points. The path

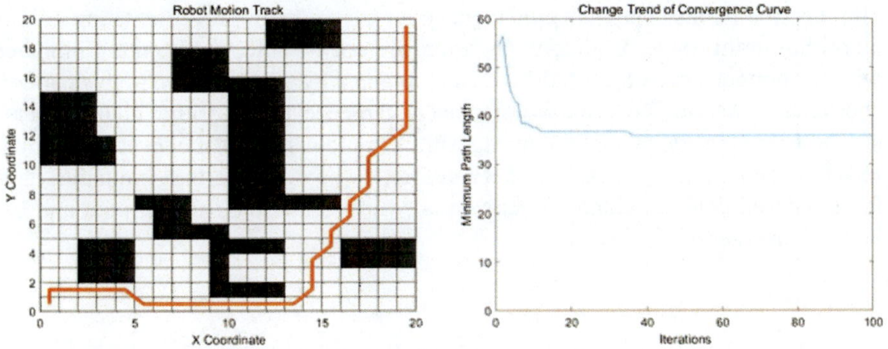

Fig. 2 Simulation results of traditional elite ant colony algorithm in 20 × 20 grid map

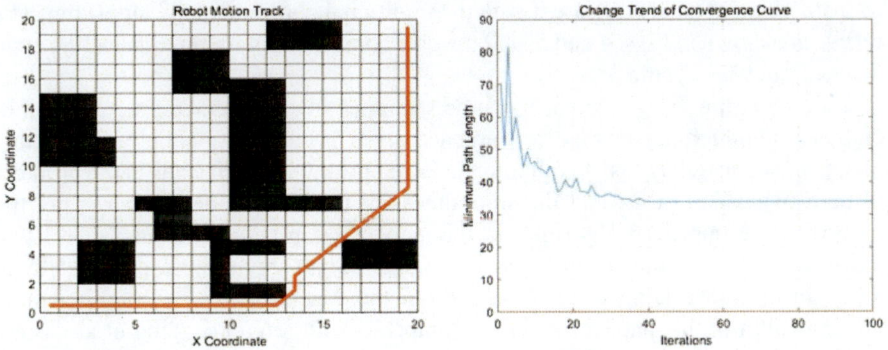

Fig. 3 Simulation results of Algorithm in this paper in 20 × 20 grid map

Table 1 Data comparison of three algorithms of 20 × 20 grid map

Evaluating indicator	Traditional elite ant colony algorithm	Reference (Liu and Liang 2021) algorithm	The algorithm in this paper
	$\alpha = 1$ $\beta = 5$ $\rho = 0.2$ $e = Q$		$\alpha(k) = 0.6 + 1 \times k/(K + k)$ $\beta(k) = 3.5 + 3.5 \times k/(K + k)$ $\rho(k) = 0.15 + 0.3 \times k/(K + k)$ $e(k) = [0.3 + 0.6 \times k/(K + k)] \times Q$
Path convergence length(cm)	35.899	39.21	33.899
Convergence times	36	52	38
Number of inflection points	13	10	4

length of the algorithm in reference (Liu and Liang 2021) is 39.21cm, and there are 10 inflection points. The path convergence length of the improved elite ant colony algorithm in this paper is 33.899, and there are only 4 inflection points. Compared with the previous two algorithms, the path obtained by the algorithm in this paper is shorter and smoother. In terms of iterative convergence, the algorithm in this paper is similar to the traditional ant colony algorithm, but lower than the convergence times in reference (Liu and Liang 2021). Therefore, compared with the simulation data, the improved elite ant colony algorithm is easier to find the shortest path and has better convergence.

5.2 30 × 30 Grid Map Simulation Experiment

Expand the scale of grid map, use the 30 × 30 grid map of reference (Jing et al. 2019) for path planning, and compare it with the traditional elite ant colony algorithm. The results are shown in Figs. 4 and 5, and the data comparison with reference (Jing et al. 2019) is shown in Table 3.

By comparing Figs. 4 and 5, it can be seen that when the environment becomes complex, although the average convergence times of the traditional elite ant colony algorithm are less, it is easy to fall into the local optimum, which leads to the problem of path twists and turns, and the smoothness of the path is low, which is not the case with the improved algorithm in this paper. It can be seen from Table 3 that the improved algorithm in this paper can still obtain the optimal path length of 42.18 in reference (Jing et al. 2019). From the aspect of average running time, the algorithm in this paper saves 55% compared with reference (Jing et al. 2019). Analysis of the reason can be attributed to the simpler algorithm in this paper; From the number of inflection points of the optimal path, the algorithm in this paper has

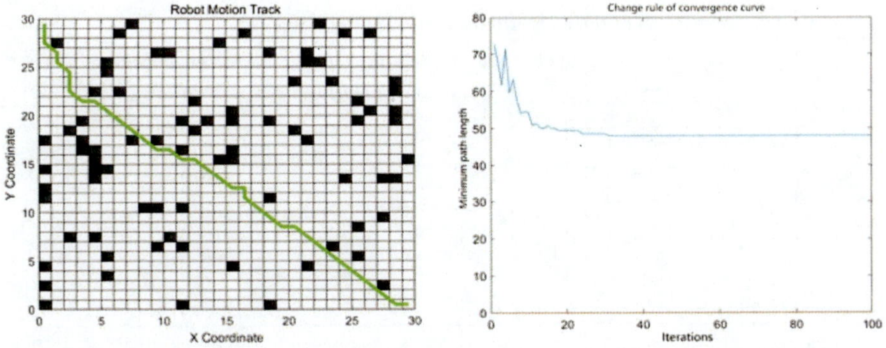

Fig. 4 Simulation results of traditional elite ant colony algorithm in 30 × 30 grid map

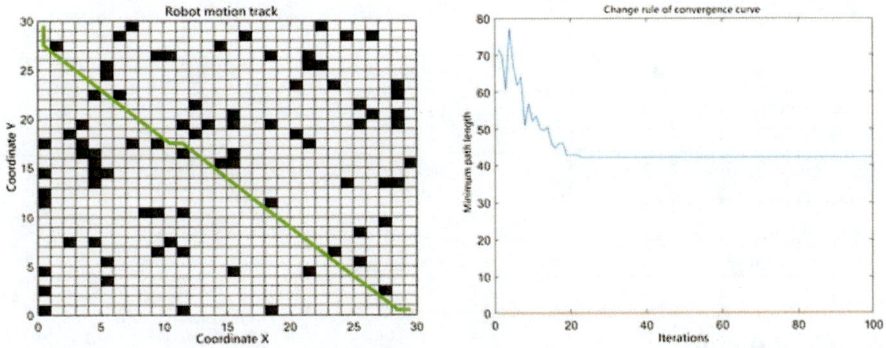

Fig. 5 Simulation results of Algorithm in this paper in 30 × 30 grid map

only 5 inflection points, less than 7 inflection points in the reference (Jing et al. 2019), and the smoothness of the path has been improved; However, the average number of iterations in this paper is more than that in reference (Jing et al. 2019). To sum up, in a complex environment, the algorithm in this paper is obviously superior to the traditional elite ant colony algorithm, with better optimization and smoothness.

Table 3 Data comparison of three algorithms of 30 × 30 grid map

Evaluating indicator	Traditional elite ant colony algorithm	Literature (Jing et al. 2019) algorithm	The algorithm in this paper
	$\alpha = 1$ $\beta = 5$ $\rho = 0.2$ $e = Q$		$\alpha(k) = 0.6 + 1 \times k/(K + k)$ $\beta(k) = 3.5 + 3.5 \times k/(K + k)$ $\rho(k) = 0.15 + 0.3 \times k/(K + k)$ $e(k) = 0.3 + 0.6 \times k/(K + k)] \times Q$
Optimal path convergence length (cm)	44.527	42.18	42.18
Worst path convergence length (cm)	53.012	42.77	43.355
Average length (cm)	47.307	42.24	42.828
Average convergence times	14.5	15.20	20.6
Average operation time (s)	22.68	46.79	21.18
Optimal path inflection points	17	7	5

6 Conclusion

The improved elite ant colony algorithm proposed in this paper with α (Information elicitation factor), β (expectation elicitation factor), ρ (Pheromone volatilization coefficient) and e (optimal path pheromone increase coefficient) are changed to dynamic adaptive parameters, which change with the number of iterations, so as to avoid the problem that traditional elite ant colony algorithm is prone to fall into local optimization. Through many comparative experiments, the results verify that the improved elite ant colony algorithm has significantly improved the path optimization ability, smoothness and robustness compared with the traditional elite ant colony algorithm.

References

Fadzli SA, Abdulkadir SI, Makhtar M, et al (2015) Robotic indoor path planning using Dijkstra's algorithm with multi-layer dictionaries. In: Int. conf. on information science and security. Seoul, 1–4

Jing Z, Yunfeng T, Guoping J et al (2019) Path planning for mobile robots based on improved ant colony algorithm. J Nanjing Univ Posts Telecommun (nat Sci Ed) 39(6):73–78

Liu X, Liang Y (2021) Research on track planning of elite ant colony algorithm with improved guidance factor. Mach Tool Hydraul 49(20):6–11

Ma X, Qian Z (2021) Research on improved ant colony algorithm in robot path planning. Comput Eng Appl 57(05):210–215

Wang X, Zhang R, Guochang G (2003) Robot global path planning based on potential field grid method. J Harbin Eng Univ 24(2):170–174

Wang J, Wu X, Guo B (2012) Mobile robot path planning based on improved particle swarm optimization algorithm. Comput Eng Appl 48(15):240–244

Wang J, Wang X, Qunhong T, Sun A, Zhang X, Yuan L (2021a) Path planning of mobile robot based on improved fuzzy adaptive genetic algorithm. Mach Tool Hydraul 49(23):18–23

Wang J, Yang Y, Li L (2021b) Mobile robot path planning based on improved depth reinforcement learning. Electron Meas Technol 44(22):19–24

Xu L, Fu W, Jiang W et al (2021) Mobile robot path planning based on 16 direction 24 neighborhood improved ant colony algorithm. Control Decis 36(5):1137–1146

Zhu D, Mingzhong Y (2010) Overview of Mobile Robot Path Planning Technology. Control Decis 25(7):961–967

Petri Net-Based Prediction of the Maintenance Time of a Special Vehicle Model

Haibing Zhang, Shiying Tian, Xinhai Shao, Aihua Zhu, and Hongkai Wang

Abstract This paper focuses on the fault maintenance of a special vehicle model. A modified Petri net is used to analyze the fault maintenance with a quantitative analytical method. Additionally, a maintenance model is established, and examples are discussed to predict the maintenance time, which serves as a basis for inspection and maintenance during the use of the special vehicle model. Currently, the failure to accurately identify vehicle faults leads to the imperfect allocation of maintenance resources and excessive spare parts. In order to achieve precise maintenance and offer pertinent support, several support units are developed and flexibly configured based on specific requirements, so that the vehicle can be restored promptly. Support units are classified into the accompanying maintenance support team, the outdoor maintenance center, and the comprehensive support base, according to the actual situation. Moreover, a modified Petri net is employed to perform a quantitative analysis of the maintenance time.

1 The Modified Petri Net

1.1 The Basic Petri Net (Gan et al. 2005; Xue'er et al. 2022; Zhang et al. 2017)

The Petri net was first proposed by the German mathematician and computer scientist Dr. Carl Adam Petri, and was mainly used to describe the information flow model with a net structure. It is a graphic tool of mathematical modeling used to describe and

H. Zhang · S. Tian · A. Zhu · H. Wang (✉)
Nanjing Campus of Army Academy of Artillery and Air Defense, Nanjing, China
e-mail: wanghk0591@mails.jlu.edu.cn

X. Shao
The First Military Representative Office of Army Equipment Department in Harbin, , Harbin, China
e-mail: sxhai86@sina.com

S. Yadav et al. (eds.), *Proceedings of 2nd International Conference on Artificial Intelligence, Robotics, and Communication*, Lecture Notes in Electrical Engineering 1063, https://doi.org/10.1007/978-981-99-4554-2_6

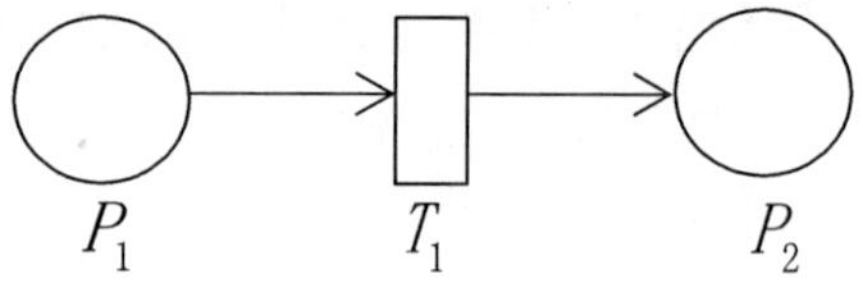

Fig. 1 The basic Petri net

analyze information processing systems that are concurrent, asynchronous, parallel, uncertain, and distributed.

The structure of the basic Petri net is a directed graph marked with operational directions, which can mainly be classified into places, transitions, tokens, and arcs. Specifically, places are generally represented by hollow circles and used to describe the states and conditions of all components of the dynamic system described. Transitions are generally depicted as hollow rectangles and used to indicate an event that changes the state of the place. For instance, a transition can describe the time of change from the part to be maintained to the part maintained. Tokens are generally represented by solid dots and mainly transited between places. A specific process of transition demonstrates changes in the state of a system. The position of a token means the state of the system after change. A place can have multiple tokens. The number of tokens in each place indicates the number of consensuses. Arcs, depicted as arrows, run from a place to a transition or vice versa. An arc displays the relationship between a part of a system with an event. See Fig. 1 for a simple structure of the Petri net.

In accordance with the relevant definitions of the Petri net, the structure can be represented by a triple, that is

$$N = (P, T, R)$$

wherein, P stands for the set of places, T is the set of transitions, and R is the set of relationships between places and transitions. There is at least one element in the sets, P and T. Tokens in a place are also known as marks and are expressed by A. If the number of initial marks, A_0, in the initial place, is ≥ 1, T_1 that connects with P_1 will have the right to occur. After the transition occurs, the system will generate a new mark and moves on according to the net structure. As transitions occur continuously, marks will change continuously.

1.2 *The Modified Petri Net*

Due to the limits of operating conditions, the vehicle needs to be constantly maintained during use. The maintenance process includes two maintenance activities: the activity from the part to be maintained to the part maintained and the activity from the part maintained to the part to be restored. During this process, the maintenance

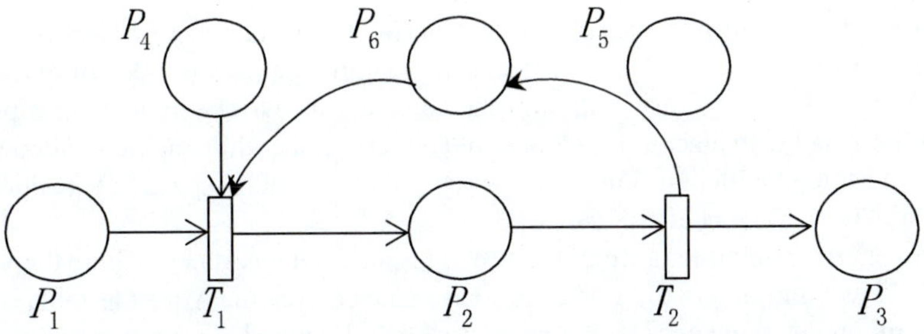

Fig. 2 Petri net of the maintenance process. P_1: the part to be maintained, P_2: the part maintained, P_3: the part restored, P_4 and P_5: tonsumed spared parts, P_6: maintenance personnel and tools, T_1 and T_2: maintenance activities

personnel use tools and consume spare parts. See Fig. 2 for the Petri net of the maintenance process.

According to the above definitions, the Petri net of maintenance describes a basic maintenance process or activity. Places describe the maintenance feasibility of a certain part to be maintained and the preparation of maintenance resources. A maintenance task or step is modeled in line with the maintenance manual. The transition contains the predicate of the execution of the maintenance task and activity. The principle of the modification of a Petri net is the method that describes the maintenance process with the maintenance network. Formalized definitions are as follows (Li and Fan 2021):

$$N = (P, T, F, W, M_0)$$

wherein:

$P = (P_1, P_2, \ldots, P_{\mathrm{m}},)$ is the set of finite places;
$T = (T_1, T_2, \ldots, T_{\mathrm{m}},)$ means the set of finite transitions;
$P \cap T = \varphi$ and $P \cup T \neq \varphi$;
$F = (P \times T) \cup (T \times P)$ is the set of directed arcs;
W: $W = F \to \{0, 1, 2 \ldots\}$ stands for the weight function of arcs;
M_0: $M_0 = P \to \{0, 1, 2 \cdots\}$ refers to the initial mark of the network.

2 Modeling of the Maintenance Process of a Broken Vehicle

In line with the description of the modified Petri net, five elements of maintenance, that is, places, transitions, directed arcs, marks, and the average implementation speed of transitions, are used to model the maintenance process of a broken vehicle. Specifically, P (place) stands for the states of the broken vehicle, such as determined to be maintained, wait to be maintained, and being restored. T (transition) means the processing and transmission of maintenance information of the broken vehicle.

F (directed arc) refers to the relationship between the states and transitions of the broken vehicle. $\lambda_i = (1, 2, 3 \ldots)$ is the average implementation speed of transitions. M (mark) indicates the distribution of the vehicle's states. The maintenance process can be depicted with places, transitions, and directed arcs, while maintenance analysis can be described with marks and the average implementation speed of transitions (Li and Fan 2021; Zhan et al. 2016).

The operating maintenance of the vehicle is a complete process where the vehicle breaks down and enters the operating maintenance system. After the fault is identified, the maintenance grade can be analyzed and defined. There are three maintenance grades in the operating maintenance system: maintenance by the accompanying maintenance support team, maintenance by the outdoor maintenance center, and maintenance by the comprehensive support base (Mu et al. 2015; Li and Zhong 2018).

Each grade of the Petri net model of the maintenance process is a complicated, internal maintenance process, which can be described with a Petri subnet, as shown in Fig. 3.

p_2, p_3, and p_4 mean the probabilities of the broken vehicle being rated as the three maintenance grades. The value is determined based on the actual situation. Tables 1 and 2 present the meanings of places and transitions in the Petri net of the operating maintenance system.

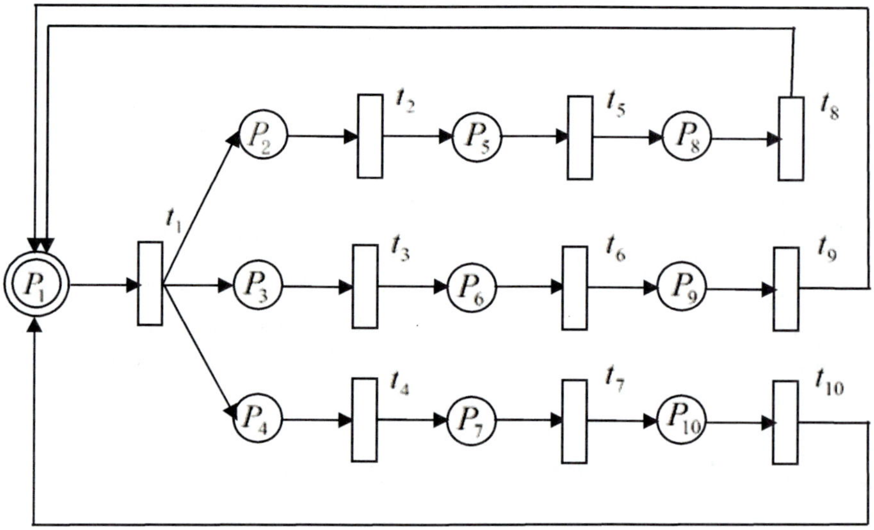

Fig. 3 Petri net of vehicle maintenance

Table 1 Meanings of places in the model

Place	Meaning
p_1	A vehicle in good condition
p_2	A broken vehicle determined to be maintained by the accompanying support team
p_3	A broken vehicle determined to be maintained by the outdoor maintenance center
p_4	A broken vehicle determined to be maintained by the comprehensive support base
p_5	A broken vehicle waiting to be maintained by the accompanying support team
p_6	A broken vehicle waiting to be maintained by the outdoor maintenance center
p_7	A broken vehicle waiting to be maintained by the comprehensive support base
p_8	A broken vehicle that is being maintained by the accompanying support team
p_9	A broken vehicle that is being maintained by the outdoor maintenance center
p_{10}	A broken vehicle that is being maintained by the comprehensive support base

Table 2 Meanings of transitions in the model

Transition	Meaning
t_1	Determination of the maintenance grade of the broken vehicle
t_2	Delivery of the vehicle to be maintained to the accompanying support team
t_3	Delivery of the vehicle to be maintained to the outdoor maintenance center
t_4	Delivery of the vehicle to be maintained to the comprehensive support base
t_5	The accompanying support team is maintaining the broken vehicle
t_6	The outdoor maintenance center is maintaining the broken vehicle
t_7	The comprehensive support base is maintaining the broken vehicle
t_8	The accompanying support team is restoring the broken vehicle
t_9	The outdoor maintenance center is restoring the broken vehicle
t_{10}	The comprehensive support base is restoring the broken vehicle

3 Determination of the Maintenance Time of a Vehicle Failure

The average implementation time, T_i, of the operating maintenance system can be calculated via the formula, $N_i = \lambda_i T_i$, wherein, N_i stands for the number of subsystems in the Petri net in the steady state, while λ_i means the number of vehicles that enter a subsystem in the unit time. The time for each subsystem to complete maintenance varies by maintenance grade (Gan et al. 2005) (Zhuang 2019; Luo et al. 2015).

The parameter $P\{(P_i)\}$ in the model of the operating maintenance system refers to the probability of transition to the state. Its calculation combines the following factors: In the preparation phase for maintenance, comprehensive factors, such as the distance between maintenance units and the maintenance strength of each maintenance unit, will be considered after the maintenance grade is determined. In practice, not all broken vehicles can enter the maintenance unit as scheduled. As a result, $P\{(P_2)\} = 0.75$, $P\{(P_3)\} = 0.54$, and $P\{(P_4)\} = 0.41$. In the phase of waiting to be maintained, the maintenance strength of each support unit and the configuration of parts under maintenance will be taken into account. In terms of the three grades of maintenance units, the probability of vehicles waiting to be maintained is determined as $P\{(P_5)\} = 0.65$, $P\{(P_6)\} = 0.63$, and $P\{(P_7)\} = 0.58$. In the maintenance phase, the maintenance strength of each maintenance unit will be considered, including the quality of maintenance personnel and the availability of maintenance tools. Hence, $P\{(P_8)\} = 0.75$, $P\{(P_9)\} = 0.84$, and $P\{(P_{10})\} = 0.93$.

Suppose six vehicles were damaged to different extents after operating for a period of time. Upon damage assessment, the number of vehicles to be maintained by the accompanying support team, the outdoor maintenance center, and the comprehensive support base are five, three, and two, respectively.

The number of vehicles that can be restored within the unit time by each subsystem is:

$$\lambda_1 = 5 \times P\{(P_2)\} \times P\{(P_5)\} \times P\{(P_8)\} = 1.8(\text{sets})$$

$$\lambda_2 = 3 \times P\{(P_3)\} \times P\{(P_6)\} \times P\{(P_9)\} = 0.8(\text{sets})$$

$$\lambda_3 = 2 \times P\{(P_4)\} \times P\{(P_7)\} \times P\{(P_{10})\} = 0.2(\text{sets})$$

The average execution time, T1, T2, and T3, of the three maintenance grades are: $T_1 T_2 T_3$.

$$T_1 = N_1/\lambda_1 = 2.8(\text{hours});$$

$$T_2 = N_2/\lambda_2 = 3.8(\text{hours});$$

$$T_3 = N_3/\lambda_3 = 10(\text{hours}).$$

4 Strategies for Maintenance Support

The analytical method proposed in this paper is an intuitive and graphic quantitative method. The modified Petri net is advantageous in the determination of maintenance resources, such as maintenance time. From the calculation process of the example given in this paper, the average execution time of different maintenance grades meets the actual use of vehicles, which implies that the allocation of resources of maintenance grades can be determined based on maintenance time so as to adopt maintenance strategies and raise work efficiency reasonably. Maintenance support should adopt the following strategies:

First, the maintenance force in the system should be given full play. Therefore, the accompanying support team, the outdoor maintenance center, and the comprehensive support base in the system should develop targeted services, reasonably arrange training according to tasks, and strictly create complicated use scenarios. Meanwhile, they should use flexible and diverse support methods to strengthen support efficiency.

Second, the hierarchy of supplies should be streamlined to save time and improve efficiency. The technical support force should be organized to comprehensively and conscientiously inspect vehicles before use, especially key and quick-wear parts. Furthermore, preparation should be fully made, and pertinent supplies and equipment should be claimed based on experience and the model budget.

Third, the local maintenance force should be combined to achieve joint support. Local technical personnel, supplies, and equipment should be fully used to expand the institution, add manpower, and offer integrated support to ensure that parts are always in good technical condition.

References

Gan MZ, Kang JS, Gao Q (2005) Military equipment maintenance engineering. National Defense Industry Press, Beijing, pp 310–315

Li JQ, Fan YS (2021) Petri net-based workflow model performance analysis method. J Comput Appl 21:4–9

Li T, Zhong SS (2018) Workflow model with colored timed petri net and its performance analysis. J Comput-Aided Des Comput Graph 18:824–831

Luo S, Zhang C, Zhou M, Pan L (2015) Researches on penetration attacking model based on timed Petri Nets. Beijing Inst Technol 35(01):92–96

Mu H, Yu J, Liu L (2015) Optimization of traffic signal actuated control at isolated intersection based on hybrid Petri net. J Central South Univ (Sci Technol) 46(07):2727–2734

Xue'er D, Jun N, Kaile Z, Xinyi M (2022) Code search method based on the reachability analysis of petri nets. J Comput Res Develop 59(01):236–250

Zhan B, Shihong M, Yanbin S, Yiping C, Yunhe H (2016) Fault diagnosis model based on improved fuzzy petri net considering time constraints. Trans China Electrotech Soc 31(23):107–115

Zhang Y, Zhang H, Wang H (2017) Security analysis on railway network time protocol based on colored petri nets. J China Railway Soc 39(10):82–88

Zhuang BY (2019) Operation control of the Petri Net for resource consumption based on reachable graph analysis. Xidian University, Xi'an, pp 7–11

Research on Modern Intelligent Sofa Design for Solitary Youth

Wenying Dong and Yanting Long

Abstract Under the background of globalization and social transformation, a large number of solitary youth have appeared in Chinese cities (XinHaiyan, Youth Explor, 1–12 (9 Nov 2022)), and solitary life has become a new life trend. Modern cities provide many young people who strive to create their own values with opportunities to develop themselves and realize their own value, which means that most young people choose a new way of life. Research and analysis of the living needs of modern solitary youth, combined with the current status of smart sofa design, explores the modern smart sofa design methods for solitary youth, provides more human care and emotional comfort for a large number of solitary youth in the city, and provides more inspiration and reference for related design.

1 Introduction

For young people living alone, time has become a luxury that cannot be wasted at will. Those who are efficient will prefer comfortable and convenient furniture. At the same time, they will also have higher requirements for furniture, such as health, fashion, personality, art and other additional requirements. Due to the rapid development of artificial intelligence, smart furniture has been continuously updated, and the use of scenes is also expanding. Among them, the sofa is a must for young people living alone and is also the main activity area in the room. Therefore, in modern cities, it is very necessary to create a harmonious environment to make living alone interesting and relaxed. It is necessary to design a modern smart sofa for "young people living alone" to meet their current situation and needs.

W. Dong (✉) · Y. Long
School of Forestry, Department of Product Design, Sichuan Agricultural University, Chengdu, Sichuan, China
e-mail: doreen0656@126.com

W. Dong
School of Furniture and Art Design, Central South University of Forestry and Technology, Changsha, Hunan, China

S. Yadav et al. (eds.), *Proceedings of 2nd International Conference on Artificial Intelligence, Robotics, and Communication*, Lecture Notes in Electrical Engineering 1063, https://doi.org/10.1007/978-981-99-4554-2_7

2 Socialization Characteristics and Needs of Young People Living Alone

The research object of this paper is the 14–35 year-old youth pointed out in the "Medium and Long-term Youth Development Plan (2016–2025)", mainly the solitary youth after they have participated in the work. Urban development cannot be separated from the integration of a large number of young people, and solitary living is a transitional stage in the development of young people. Professor Lin Zuoxin of Beijing Forestry University once made a detailed survey on the furniture consumer market. The research shows that eight stages in a person's life are the main period of furniture purchase, and the youth stage is one of the stages of furniture purchase. Therefore, the design of smart sofas in furniture for young people living alone has a certain market and needs.

2.1 *Socialization Characteristics*

Separated from the collective life of the campus and family units, they began to live alone and face the society, and were full of struggle and rising passion in their work, which belonged to the accumulation of society, and were full of struggle and rising passion in their work, which belonged to the accumulation of resources. At the same time, they are the earliest generation in society to have access to the Internet and intelligent products. They often receive fresh knowledge and cutting-edge technology. Therefore, it is easy to accept and enjoy products with high and new technology. With a certain income, they are also willing to invest in their own preferences and moods and are more willing to purchase and set up intelligent products that they really like. Secondly, after a day's work, dragging your exhausted body home, what you need most is a comfortable rest, your own time and some fun in life. If you spend a long time alone after work, you will inevitably experience psychological changes. Therefore, if frequently used furniture is comfortable and interesting, it will be more in line with their current situation and immediate needs. A person living alone is also their shelter in the city and their warm home.

2.2 *Needs of Young People Living Alone*

According to the eighth-order Maslow's hierarchy of needs (Maslow 1970), the living room is a social location that meets the human needs for belonging and love. Among 100 young people living alone, the furniture most frequently used in their living room was investigated (see Fig. 1). As you can see from the chart, when you return home to rest, the sofa in your living room is the most frequently used, followed by the tea table. The explanation given by the solitary youth is that the sofa can accompany the

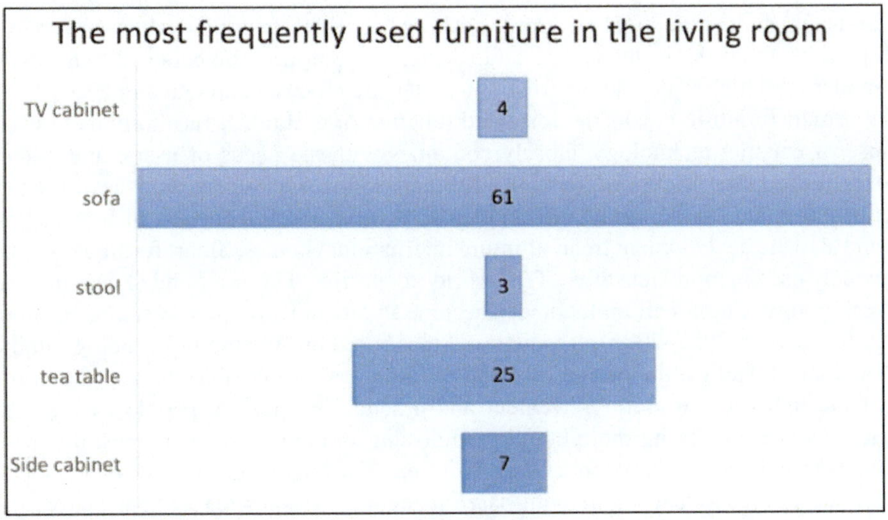

Fig. 1 Which furniture is most frequently used in the living room (Author Self-Drawing)

user through many solitary times, such as relaxation, games, chatting, reading, etc. Therefore, in the life of solitary youth, a sofa is an important furniture to accompany them for a long time. At the same time, the emphasis on sofa design also meets their needs for respect and aesthetics.

3 Current Status and Key Points of Modern Smart Sofa Design

"Modernization" is interpreted in the Chinese dictionary as "bringing about the advanced level of modern science and technology".With the development of artificial intelligence and the Internet of Things, we can see the advanced level of modern science and technology, and bring us into a new and modern world. Nowadays, smart homes and smart home systems are the main household markets consumed by young people. In the top 10 of the "Wireless Smart Home System (China market)" (see Fig. 2), Apple, Green rice and Xiaomi form a "tripod". Then the brand can see that most of the brands in China, indicating that the domestic market has made a lot of investment in smart homes, and there is a lot of room for growth in the future. Among them, smart home mainly has small products in the consumer market, and the most smart products are small appliances in the home. Among them, smart Bluetooth speakers are the most, such as EZVALO, Apple, Elvis Presley and other brands. In smart furniture, the design is mainly focused on smart beds and smart bedside tables in the bedroom. Most of them add some scientific and technological functions to traditional furniture, such as smart beds that can play music, sofas with automatic massage, multi-functional electronic shaking beds, elevated office tables and chairs, etc. Lack of aesthetic and comfortable design, cannot meet the personality needs of modern young people and the trust of a better life (Jiazhen and Ouyang 2023).

Smart furniture breaks the form and function of traditional furniture, makes full use of existing technology, closely combines with the needs of users, and makes the future home life more humanized. Smart sofa is one of the branches of smart furniture, and the design of the smart sofa is the perfect deduction of human life, and also the development trend of future international homes. Smart furniture will be widely used in more than 80% of the world's countries. This is a trend of demand and technology. China will undoubtedly become the main battlefield of smart furniture (CIC Advisor 2022–2026). It is believed that based on Bluetooth technology in the Internet of Things, the design of furniture products will be able to meet the needs of the times, bring more convenient and practical services to people's lives, and at the same time bring more humanized design. In modern smart homes, the main connections used are Bluetooth, WiFi, NFC, etc. The Internet of Things has developed rapidly in recent years. Bluetooth-based technology has a wide audience, covering almost all smart applications. Bluetooth standard has a large proportion in the Internet of Things application market due to its intelligence, low power consumption, high connection speed and low cost. Therefore, the satisfaction of Bluetooth use in life was investigated (see Fig. 3). 100 young people were randomly selected, and 94% of them were very satisfied and basically satisfied. The reason is that they are easy to use, fast and do not need to repeat the connection, you can also personalize the product without downloading a lot of apps. In 2020, the scale of China's artificial intelligence industry reached 303.1 billion yuan, Artificial Intelligence Industry was also selected

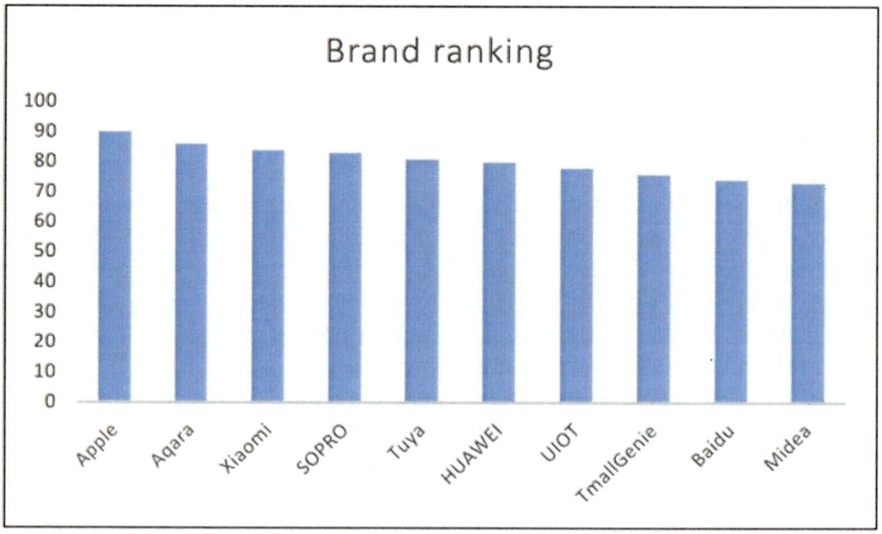

Fig. 2 Wireless Smart Home System (China market) leaderboard (from network, drawn by the author)

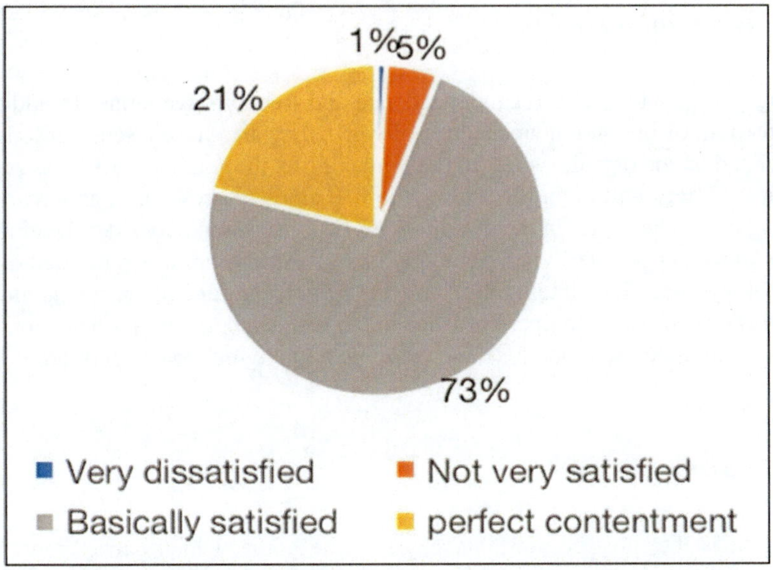

Fig. 3 Bluetooth use satisfaction survey in modern youth life (Author Self-Drawing)

as the top ten investment hotspots of investment advisors in 2022 (OFweek Smart home network 2016). With the collaborative development of artificial intelligence and the Internet of Things, China will have a broader and more positive development in the smart furniture market.

4 Modern Intelligent Sofa Design Strategy for Solitary Youth

4.1 Individualization Design

Each group is different, especially the youth living alone has its distinctive characteristics. Therefore, in the design of intelligent sofa, it is necessary to distinguish from traditional furniture design, not simple function stacking or material collage. Some materials of the sofa can best be changed and matched according to the personal preferences of young people living alone so that the material design of the smart sofa can meet their personalized needs. In a word, the personalized design of smart sofa, mainly emphasizes the design based on "people", respects the characteristics of young people living alone, and meets their aesthetic needs for fashion, cutting-edge and artistry.

4.2 Emotional Design

Young people who yearn for independence and freedom live alone. In addition to the freedom of life and independence of work, they also have a sense of loneliness stimulated by the outside world. In the prime of life, they endure loneliness and loss. A warm society will not allow those who live alone to grow up in an environment lacking care. Therefore, in the design of the sofa, we should consider the emotional needs of young people living alone, give them warm care, integrate the advanced technology of artificial intelligence, and use the sensing function of temperature and sound. The sensing of temperature and sound can detect the user's behavior, adjust the most suitable use mode for them, and make the living alone full of temperature.

4.3 Funny Design

The sofa is the most comfortable place in the home, which carries most of the entertainment and leisure activities of young people living alone. Therefore, the design of the smart sofa should take into account the behavior and activities of solitary people, add points for their entertainment, and add more interest to the solitary life. For example, light adjustment is designed in the smart sofa, and different lighting effects are designed according to their different entertainment behaviors to become the background wall of their life, making the monotonous life of living alone colorful and satisfying their spiritual needs.

5 Design Practice of "Music Note Smart Sofa"

5.1 Rhino Modeling with Keyshot Rendering

According to the design needs of young people living alone, a "musical note smart sofa" was designed. The rhinoceros drawing software was used in the design process. After drawing the sofa, adjust it to wireframe mode, and you can clearly see the positions of the three-layer light strip, wireless charging and Bluetooth (see Fig. 4). In the drawing interface of the rhinoceros, you can clearly see the three views of the sofa in wireframe mode and shading mode (see Fig. 4). Finally, pour the design drawing drawn by the rhinoceros into the keyshot software for material rendering, and you can see the three views of the rendered sofa (see Fig. 5).

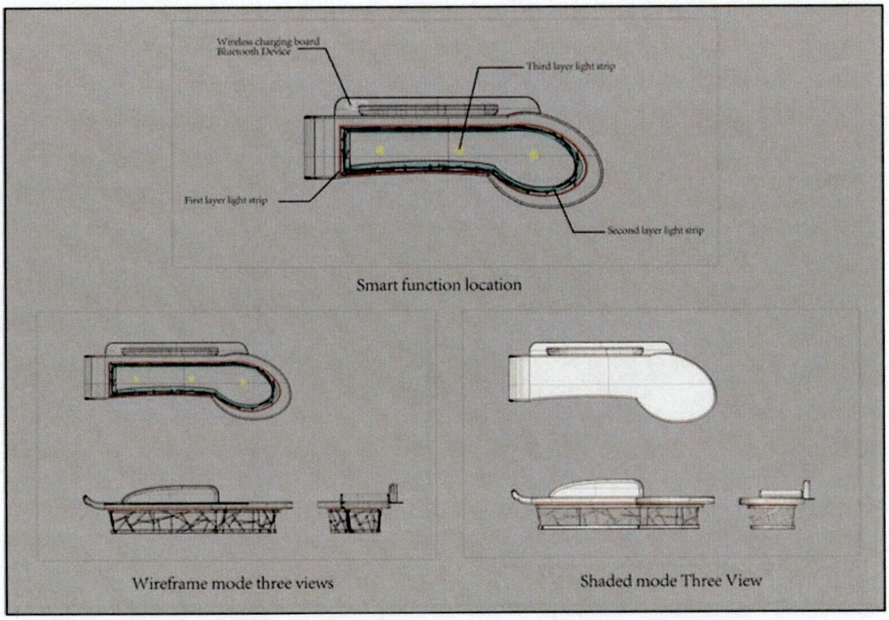

Fig. 4 Smart Function Location, Wireframe mode and shading mode three views

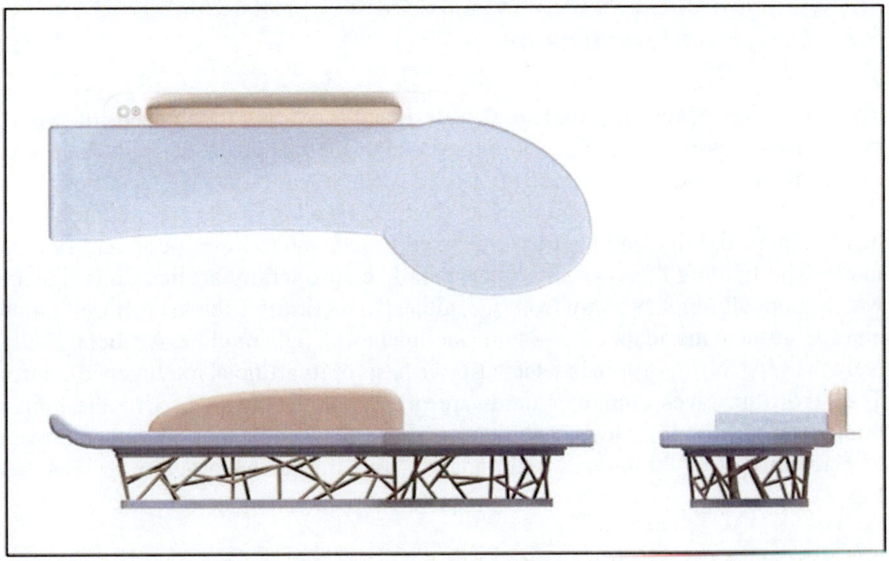

Fig. 5 Three views of sofa

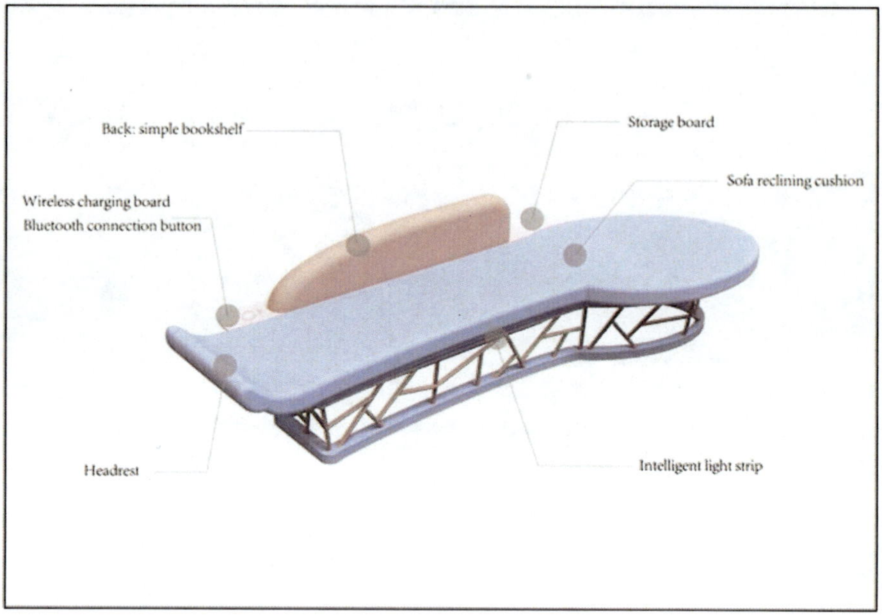

Fig. 6 Function of sofa

5.2 *Design and Application*

The function of "Musical smart sofa" is designed according to the needs of "young people living alone" (see Fig. 6), among which the design of the light belt is an important function of "Musical smart sofa", which can be connected through the Bluetooth of mobile phone and then set the light (see Fig. 7). There are three kinds of intelligent lighting modes for sofa: fixed mode, self setting mode and random mode. The lighting effect of this smart sofa is controlled by artificial intelligence, which gradually breaks away from the subjective control of the user. Based on the consideration of the adaptability of the intelligent sofa light module, Artificial Neural Networks (ANN) algorithm is selected as the basis of its artificial intelligence control. The algorithm gives computer hardware programmatically to part of the human thinking mode through a logical computing strategy to enable the computer hardware to perform part of the human brain analysis function under program driving (see Fig. 8).

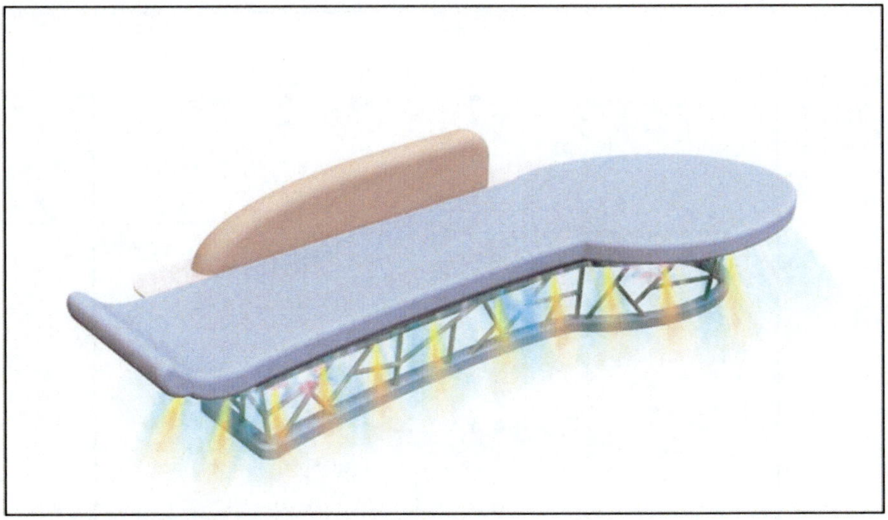

Fig. 7 Sofa lighting effect picture

In fixed mode, with a thematic lighting scheme; In the custom mode, the user can set the color of the light by himself; In random mode, artificial intelligence-based technology that analyzes the user's mood, behavior, sound, temperature, and so on, providing automatic light selection when no selection is made. When playing games, what users need is an ambience to achieve the authenticity of the game, and the lights will change according to the sound of the game; When you sleep, the light band at the bottom of the sofa becomes soft and quiet, accompanied by a drop in body temperature and sound of breathing while sleeping; While reading, the surrounding light can also be adjusted by the user himself. After a long time of use, the sofa slowly adapts and feels, forming memory, and learning itself according to the user's body, habits, mood and behavior, so that the user no longer needs to control when sitting on the sofa, so that the user can feel the care of the sofa when they return home.

Therefore, the design of "Music Intelligent Sofa" meets the personalized, emotional and interesting design needs of the solitary youth. Even in the solitary life, you can feel the sense of belonging and the care of human design brought by "Music Sofa".

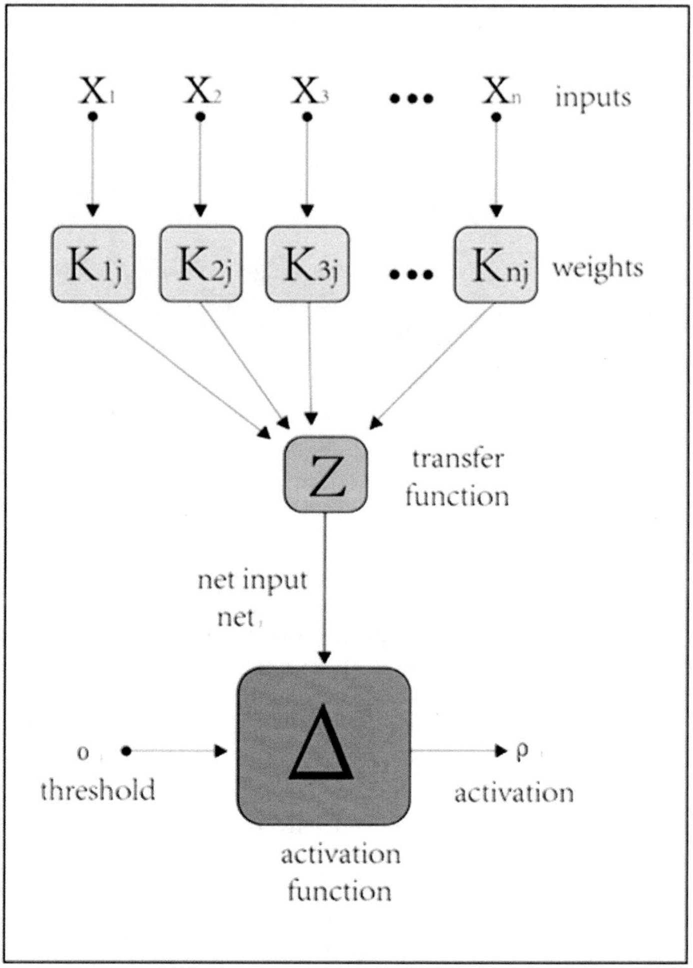

Fig. 8 Artificial neural network (Author Self-Drawing)

6 Conclusion

In this diversified society, furniture design for the needs of young people living alone is also a reflection of a harmonious society. The rapid development of modern science and technology is also aimed at serving the people. Therefore, it is very necessary and feasible to design sofas that are suitable for the real needs of young people living alone by combining the current hot AI and Internet of Things technologies. It is believed that with the continuous updating of science and technology, more furniture designs with temperature will emerge in the consumer market for young people living alone.

References

CIC Advisor. 2022–2026 China AI industry deep research and investment prospect forecast report (volumes I, II and II)

Jiazhen H, Ouyang H (2023) Research on multifunctional furniture design. School of Home and Industrial Design, Nanjing Forestry University, p 210037

Maslow AH (1970) Religions, values, and peak experiences. Penguin, New York (Original work published 1966)

OFweek Smart home network, analysis on the development planning direction of intelligent furniture in China. https://smarthome.ofweek.com/.2016/6/25

XinHaiyan D (9 Nov 2022) Rebuilding the space time boundary of lonely dwelling youth in big cities. Youth Explor, 1–12

AUV Path Planning Based on Improved Ant Colony Algorithm

Yu Liu and Ai-hua Wu

Abstract In order to solve the problem of limited-energy underwater robot AUV path planning, on the premise that the underwater navigation environment and the direction of water flow are known, an improved dual population ant colony algorithm is proposed to study the shortest path problem of AUV in the state of fixed and limited navigation energy. Firstly, the underwater three-dimensional environment is simulated based on the grid method, and the energy consumption and motion model of the underwater robot is established under the influence of water flow, and then the exploration mode based on the heterogeneous dual population is studied, and the algorithm is realized by setting the dual heterogeneous population. The convergence ability and the ability to explore the shortest path are improved, and different pheromone update methods corresponding to the heterogeneous population are set to strengthen the ability of the heterogeneous population to find the shortest path under the condition of limited energy. The experimental results show that the algorithm has a significant improvement in the ability to approximate the optimal solution and improve the convergence speed compared with the comparative methods, and the average path length and the average number of iterations are correspondingly reduced, and it has a good application in the path planning of underwater robots.

1 Introduction

The ocean has very rich natural resources. But the marine environment is very complex. Underwater robots are important equipment for exploring the ocean, and have great potential in ocean observation, exploration, and operations in extreme underwater environments. Especially for deep-sea salvage, trench scientific research

Y. Liu · A. Wu (✉)
College of Information Engineering, Shanghai Maritime University, Shanghai, China
e-mail: ahwu@shmtu.edu.cn

Y. Liu
e-mail: 202130310281@stu.shmtu.edu.cn; yu_liu202308@163.com

S. Yadav et al. (eds.), *Proceedings of 2nd International Conference on Artificial Intelligence, Robotics, and Communication*, Lecture Notes in Electrical Engineering 1063, https://doi.org/10.1007/978-981-99-4554-2_8

sample sampling, etc. path planning is one of the key technologies for underwater robots to achieve salvage, sampling, patrolling and other tasks. Common path planning algorithms include genetic algorithm, ant colony algorithm, simulation Annealing algorithm, particle swarm algorithm, etc. Among them, ant colony algorithm has been widely used in path planning because of its good global optimization ability and stability.

Easy to fall into local optimal solution, slow convergence speed, etc. are common problems of traditional ant colony algorithm. In response to these problems, the reference (Sun et al. 2017) proposes a congestion factor, and adaptively adjusts the number of ants passing through the control path according to the number of iterations to accelerate the convergence of the algorithm, the reference (Deng et al. 2019) proposes a pheromone diffusion mechanism, which can make ants release pheromone at a certain point, thereby gradually affecting adjacent areas a certain range, the reference (Xu et al. 2017) proposes a new heuristic communication heterogeneous dual population ant colony optimization to balance the convergence speed and the solution's diversity in the large-scale travel salesman problem, the reference (Li and Wang 2020) sets the initial pheromone concentration based on the position information of the current feasible target point and the end point, the reference (Yan 2021) updating pheromone combined with sorting method, the reference (Zhang and Jia 2013) proposes a new method based on Octree model and improved ant colony algorithm, the reference (Ma et al. 2018) develop a nature-inspired ant colony optimization algorithm to search the optimal path.

In this paper, a multi-swarm strategy is adopted, the elite population and the common population are used to improve the diversity of the algorithm, and the competition mechanism and the water flow direction constraint are used for the elite population to improve the convergence speed of the algorithm, and the congestion threshold is set for the common population to improve the ability of the algorithm to explore the optimal solution. At the same time, in order to improve the feasibility of AUV 3D path planning and optimize the planned path, this paper adds finite energy constraints on the basis of solving the traditional underwater robot path planning. Under the motion model of the robot, and through the improved heterogeneous multi-swarm ant colony algorithm, the optimization goal of the shortest path under limited energy is achieved.

2 Underwater Robot Modeling

2.1 Environment Model Establishment

The grid method is used for environmental modeling, which is easy to implement, can simplify the calculation, and has a standardized form. It is a commonly used environmental modeling method. The exponential function is used to describe the mountains encountered during the navigation of the underwater robot. The underwater environment modeling is shown in Fig. 1.

2.2 Energy Consumption Model

When the underwater robot moves at a uniform speed, the water flow resistance and the thrust generated by the thruster are balanced. In order to simplify the calculation, this paper ignores all the energy consumption of the underwater robot except the thruster. Therefore, the energy consumption of the underwater robot can be obtained by calculating the work to overcome the water resistance, the water flow resistance is proportional to the square of the relative water flow velocity of the underwater robot, and it can be simplified that the work done by the underwater robot to overcome the water resistance during the constant speed navigation is as follows:

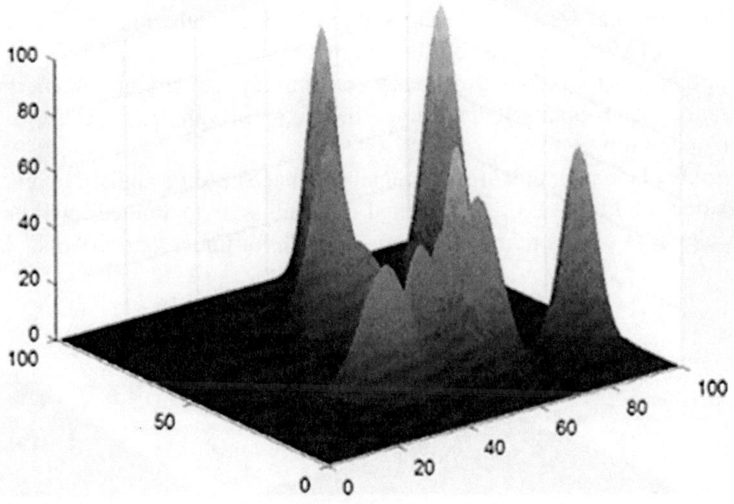

Fig. 1 Simulation of the underwater environment

$$E = \frac{Pt}{\eta} = \frac{Fvt}{\eta} = \frac{FL}{\eta} = \frac{C_d \rho v^2 SL}{2} \tag{1}$$

In the formula: L represents the path length of the AUV sailing, and η is the mechanical efficiency.

2.3 Mathematical Model

The path planning problem of the underwater robot is explained as follows: in a certain limited navigation space, plan a navigation path, and make the length J smaller under the premise of satisfying certain constraints, the planned route can use a grid. The point on the top represents $T = \{P_s, P_1, P_2, \ldots\ P_n, P_G\}$. where P_i is the selected point of each layer of the grid.

$$\min J = \sum_{i=1}^{n} S_{P_i \to P_{i+1}} \tag{2}$$

$$\forall z > z(x, y) \tag{3}$$

$$\sum_{i=1}^{n} \frac{C \rho v_i^2 SL}{2\eta} \leq W \tag{4}$$

Equation (3) for terrain threat constraints, in order to ensure the safety of underwater robots during navigation, the navigation route cannot pass through terrain obstacles and threat areas. Assuming that the planned path curve is $f(x, y, z) = 0$, where (x, y, z) represents the coordinates of any point on the path.

Equation (4) is the maximum energy consumption constraint, the energy carried by the underwater robot itself determines the maximum value of the energy consumed by the planned route.

The route planned in the three-dimensional space must be applied in practice, and the curvature and torsion of the planned route must be guaranteed without sudden changes, so as to ensure that this route is feasible for underwater robots.

3 Ant Colony Algorithm Based on Heterogeneous Double Population

3.1 Heterogeneous Dual Population Parallel Structure

For the same population of ant colony algorithm, improving the algorithm's convergence speed and improving the algorithm's ability to explore the optimal solution often cannot be achieved at the same time. The algorithm converges slowly. Therefore, using multiple colony strategies to improve the ant colony algorithm, the overall ability of the algorithm to explore the optimal solution and the convergence speed will be improved due to the complementary advantages of the two populations. The ant colony structure setting of the multi-population strategy is shown in Fig. 2.

In this paper, two heterogeneous populations are set up, namely the elite population and the common population. The elite population is responsible for maintaining the stability of the algorithm convergence, and the common population is responsible for broadening the search diversity of the ant colony algorithm.

After one round of search of the two populations is completed, the paths searched by the two populations are exchanged, and the best n ants in the ordinary population are replaced with the worst n ants in the elite population. Then proceed to the next iteration.

3.2 Probabilistic Selection Function Update

In order to accelerate the convergence speed of the elite population under the constraints of energy consumption, the influence of water flow direction is added to the probability selection function. The direction of water flow can guide the ants'

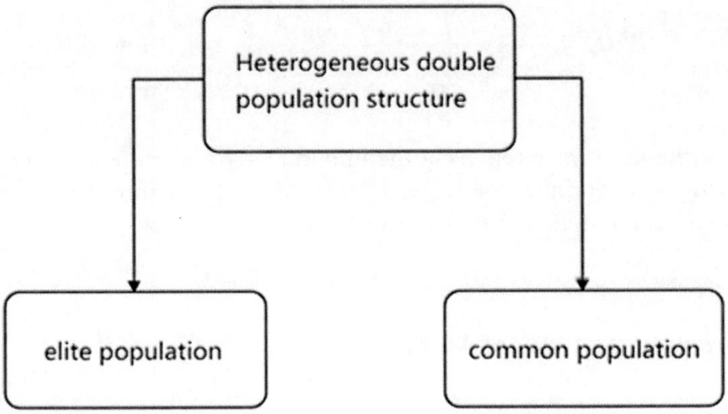

Fig. 2 Heterogeneous double population structure

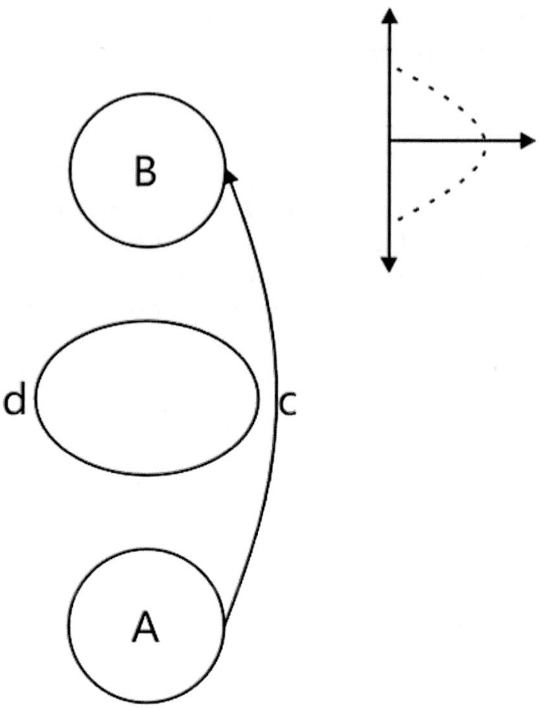

Fig. 3 Water flow diagram

path selection, which makes it easier to meet the energy constraints and the algorithm converges faster. Figure 3 shows the influence of water flow on path selection, an ant starts from point A to point B, then in the direction of water flow, the ant will tend to go in the c direction, so that the energy consumption is lower than that of the d direction, and it is easier to achieve energy constraints. The redesigned probability selection function is as follows:

$$\begin{cases} P_{ij}^{k}(t) = \frac{(\tau_{ij}(t))^{\alpha}(\eta_{ij}(t))^{\beta}(\varphi_{ij}(t))^{g}}{\sum_{s \in \text{allowed}(i)} (\tau_{is}(t))^{\alpha}(\eta_{is}(t))^{\beta}(\varphi_{is}(t))^{g}} & j \in \text{allowed } (i) \\ 0 & \text{otherwise} \end{cases} \tag{5}$$

In the formula: α, β, g represent the importance of pheromone, heuristic factor and water flow direction respectively; $P_{ij}^{k}(t)$ represents the transition probability of ant k from node i to node j at time t.

3.3 *Pheromone Update Method*

When exploring feasible paths, sub-populations tend to focus on searching the same area, resulting in a waste of feasible area resources and increasing the probability of falling into a local optimum. In order to solve this problem, the general population adopts the following rules to update the pheromone:

$$\tau_{ij}(t) = \begin{cases} \tau_{ij}\tau < \tau_{\max} \\ \tau_{\max}\tau > \tau_{\max} \end{cases} \tag{6}$$

where $\tau_{\max}$ is the maximum pheromone concentration allowed.

When the population evolves alone for a certain number of iterations, a competition mechanism is adopted for the population, and appropriate pheromone incentives or penalties are given to both winners and losers, thereby enhancing the competitive population and speeding up the convergence speed of the algorithm. For the elite population, in order to speed up the convergence speed of the solution, a penalty mechanism is adopted to update the pheromone, and the following formula is used for the pheromone generated by the optimal ant in each round of iteration:

$$\Delta\tau_{ij}^{k}(k) = \begin{cases} \frac{Q}{C_k} + P_i \text{ if component}(i, j)\text{was used by ant} \\ 0 \qquad\qquad \text{Otherwise} \end{cases} \tag{7}$$

In the formula: P_i is the pheromone reward constant of the optimal ant.

4 Experimental Simulation and Analysis

The relevant parameters in the algorithm are set as follows: the number of iterations is set to 100, the number of ants in the two populations is set to 100, $\alpha = 10$, $\beta = 1$, $g = 1$, and the pheromone volatility factor $r = 0.1$. The number of ants exchanged in each iteration is $n = 2$. Reward constant $P_i = 0.1$.

4.1 Comparative Experiment

In this section, based on the same simulated terrain, a control experiment on the energy parameter W is carried out to verify the feasibility and effectiveness of the improved ant colony algorithm proposed in this paper. All algorithms are implemented on a Windows 10 computer running at 3.2 GHz and 16 GB memory. In the simulation experiment, the navigation environment is a three-dimensional environment of 100 m × 100 m × 100 m, and is divided into 9 layers of grids along the z-axis. The step size is both is 10 m, the underwater robot starts from the coordinates (0, 0, 0) and arrives at the coordinates (100, 100, 80). The original ant colony algorithm and the algorithm in this paper were compared with the finite energy W respectively, and the value of W was changed to carry out multiple sets of experiments. The specific

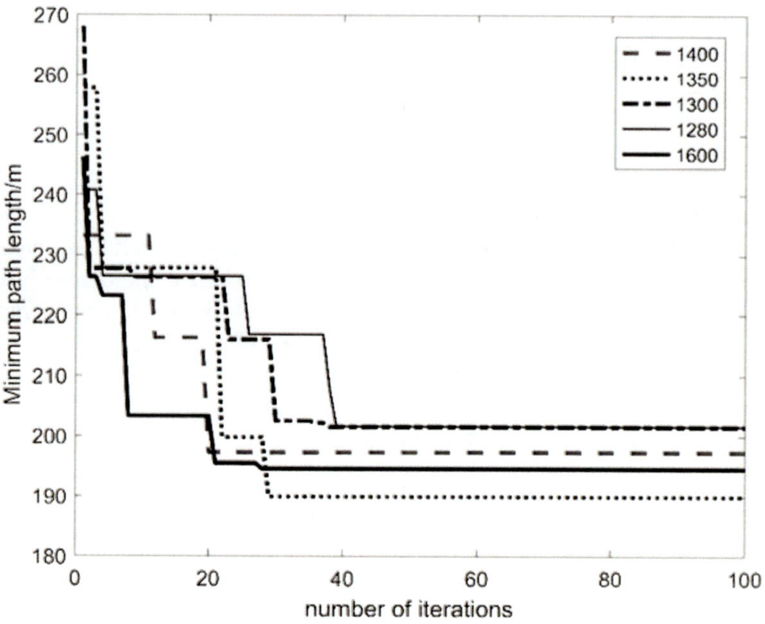

Fig. 4 The original ant colony algorithm convergence diagram

comparison is shown in Figs. 4 and 5. The performance of the algorithm under the same energy is compared.

The comparison results are shown in Figs. 4 and 5, of which Fig. 4 is the convergence curve of the original ant colony algorithm under different energy constraints. It can be seen from the convergence curve that when other conditions are the same, the restricted energy is smaller, the slower the algorithm converges, the longer the finally searched path length is. Figure 5 shows the convergence curve of the algorithm in this paper under different energy constraints. It can be seen from the figure that under the same other conditions, the limited energy The smaller the value is, the algorithm has no obvious fluctuation, all converge in the same interval, and the length of the finally found path does not change significantly. The average shortest path is reduced from 198.0 to 191.1, an average reduction of 6.9 m, and the average number of iterations is reduced from 29.8 to 20, an average reduction of 9.8.

The trajectory convergence is shown in Figs. 6 and 7, 8 and 9 are the top view of the trajectory. It can be seen from the comparison that when the energy limits are 1280 and 1300 respectively, the optimal path is reduced by 11.6 and 6.6, the number of convergence times is correspondingly reduced, and the convergence is faster, indicating that under the limited energy limit, the improved algorithm is easier to find the optimal path that satisfies the energy limit, can better avoid the blindness of search, and has also been improved in the final optimal path. The overall effect of the improved ant colony algorithm is better when the environmental variables and other conditions are the same traditional ant colony algorithm.

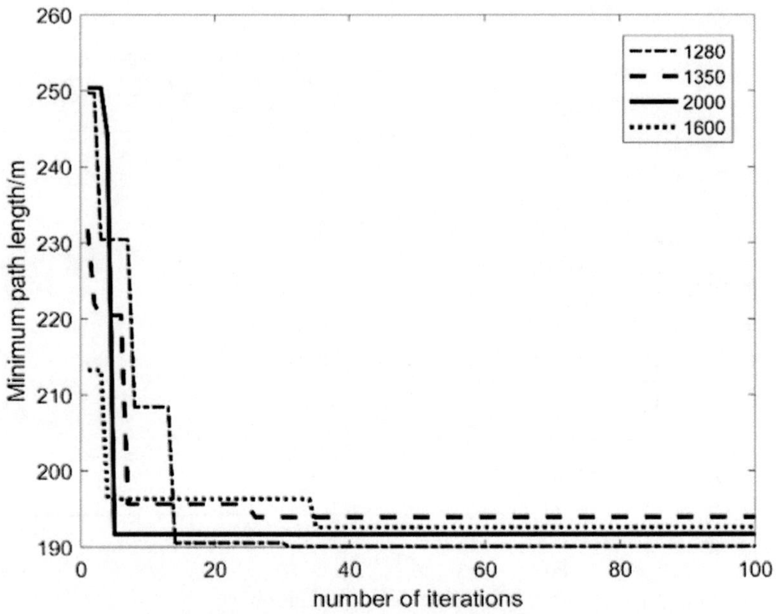

Fig. 5 Improved ant colony algorithm convergence diagram

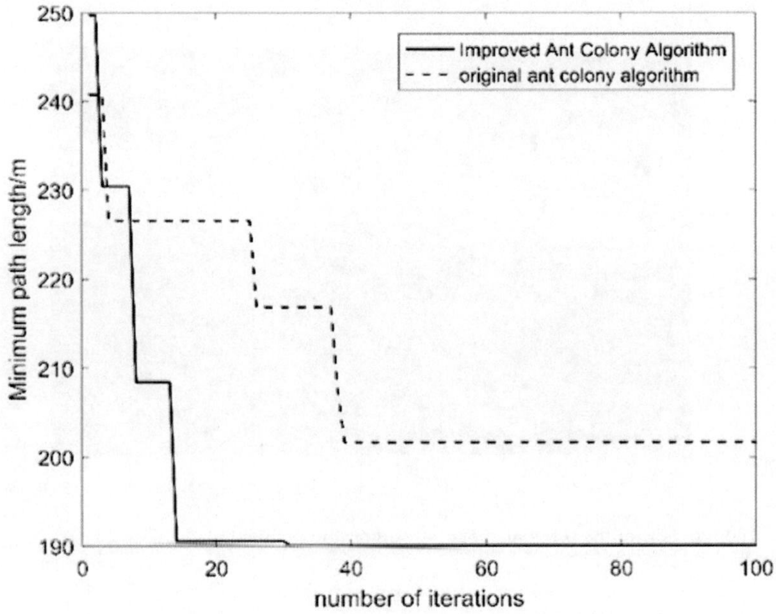

Fig. 6 Convergence comparison chart when the energy constraint is 1280

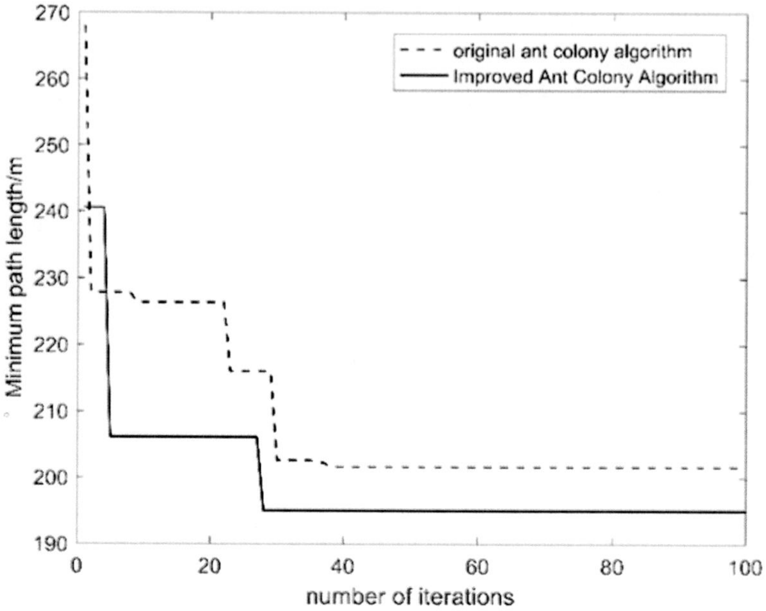

Fig. 7 Convergence comparison chart when the energy constraint is 1300

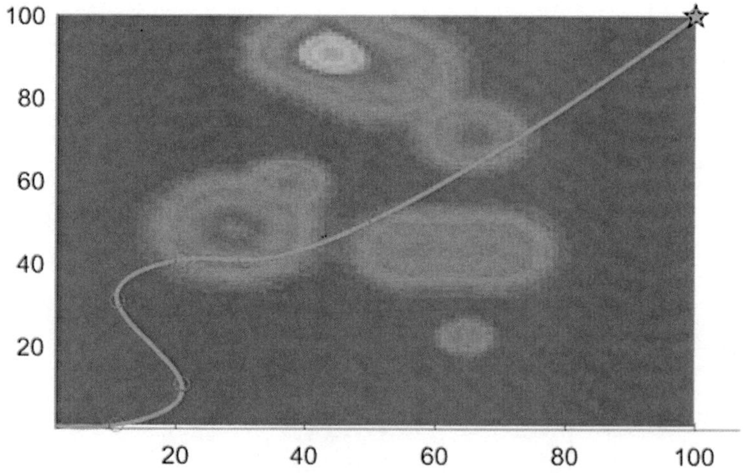

Fig. 8 The top view of the original ant colony algorithm planning path ($W = 1280$)

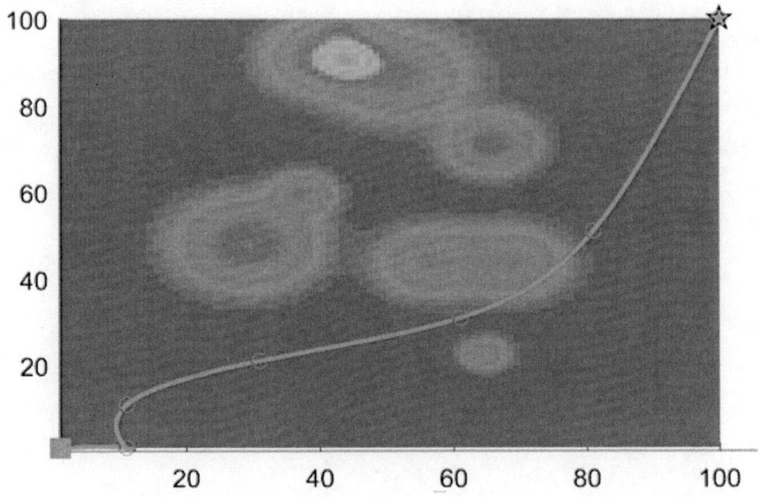

Fig. 9 Top view of improved ant colony algorithm planning path ($W = 1280$)

5 Conclusions

In this paper, on the basis of studying the application of ant colony algorithm to the path planning of underwater robots, in order to better apply it to practice, the finite energy constraints and the influence of water flow are added, and the ant colony algorithm is improved, and dual populations with different functions are added. Through the comparison experiment of the parameters, the improved algorithm in this paper is more superior, and the shortest path and the number of iterations are significantly improved.

References

Deng W, Xu J, Zhao H (2019) An Improved ant colony optimization algorithm based on hybrid strategies for scheduling problem. IEEE Access 7:20281–20292

Li J, Wang H (2020) Research on AUV path planning based on improved ant colony algorithm. In: 2020 IEEE international conference on mechatronics and automation (ICMA). IEEE

Ma YN, Gong YJ, Xiao CF, Gao Y, Zhang J (2018) Path planning for autonomous underwater vehicles: an ant colony algorithm incorporating alarm pheromone. IEEE Trans Veh Technol (1):141–154

Sun X, Zhang K, Ma M (2017) Multi-population ant colony algorithm for virtual machine deployment. IEEE Access 5:27014–27022

Xu M, You X, Liu S (2017) A novel Heuristic communication heterogeneous dual population ant colony optimization algorithm. IEEE Access 5:18506–18515

Yan S (2021) Research on path planning of AUV based on improved ant colony algorithm. In: 2021 4th international conference on artificial intelligence and big data (ICAIBD)

Zhang G, Jia H (2013) 3D path planning of AUV based on improved ant colony optimization. In: Proceedings of the 32nd Chinese Control Conference. IEEE

Serverless Computation and Related Analysis Based on Matrix Multiplication

Jiajun Gong, Haotian Sun, and Yilin Wang

Abstract This paper investigates the effect of the change in matrix dimensionality and the relationship between function communication time and execution time, as well as the relationship between memory parameters and execution time, on the efficiency and economic cost of computation when using a serverless platform for high-dimensional matrix data. AWS Lambda was used to create the functions, AWS S3 to store the data, perform the high-dimensional matrix fork multiplication and matrix convolution operations, respectively, and S3 to design the monitoring and data analysis interface, and finally, analyze the statistical data. According to the analysis, the average communication time consumption ratio decreases gradually as the number of matrix dimensions increases. The computation time decreases by increasing the number of nodes as far as concurrency allows. As the memory parameter increases, the computation time decreases, and the total cost decreases and then increases. The experimental results show that the impact of communication costs on serverless computing decreases as the size of the data increases. When the concurrency limit is not reached, there is a significant increase in computational efficiency, reducing the time consumed by individual nodes and the overall computing process. Increased memory improves computational efficiency but does not necessarily result in total cost savings.

Jiajun Gong, Haotian Sun and Yilin Wang these authors contributed equally.

J. Gong
College of Architecture and Urban Planning, Shenyang Jianzhu University, Shenyang 110000, China

H. Sun
College of Stomatology, ChongQing Medical University, Chengdu 610066, China

Y. Wang (✉)
Department of Computer Science and Technology, Tsinghua University, Beijing 100084, China
e-mail: 631301120328@mails.cqjtu.edu.cn

S. Yadav et al. (eds.), *Proceedings of 2nd International Conference on Artificial Intelligence, Robotics, and Communication*, Lecture Notes in Electrical Engineering 1063, https://doi.org/10.1007/978-981-99-4554-2_9

1 Introduction

Serverless is a cloud-native development model. The concept's core is deserialization, emphasizing architectural ideas and service models. Serverless Computing can be defined by its name—think (or care) less about servers. Developers don't need to worry about the low-level details of server management and scaling and only pay when processing requests or events. User defines serverless computing as a platform that hides server usage from developers and runs code on demand, automatically scaling and billing only when the code is running (Werner et al. 2018; Baldini et al. 2017).

This allows developers to focus less on the infrastructure and more focused on the business logic of the application, including the following: (1) Operations and Maintenance free: no need to manage server hosts or server processes. (2) Auto Scaling: automatic scale scaling and automatic configuration according to the load. The expansion range is zero to infinity. (3) Pay as you go: determine the actual cost based on usage. (4) High availability: has implied high availability. Using serverless technology to improve cost performance, it can automatically scale and contract according to the load, saving a large wave of manpower and resource costs, better managing costs and usage, and reducing the difficulty of Operations and Maintenance.

To maximize utility, suppliers will strive to improve resource use efficiency. Compared with many manufacturers now maintaining their hosts and processes, it can reduce unnecessary waste of resources. And it has better performance and higher stability, scalability, and security, which can improve efficiency, faster development, and launch, and realize the transformation from resource-based to service-oriented. As a result, developers can transition from traditional operations responsible for the maintenance of the underlying hardware to DevOps or SRE. Major cloud vendors include Alibaba Cloud, Tencent Cloud, Amazon, and Google. Successfully launched Serverless products, and this article uses Amazon Web Services. From a cloud provider's perspective, serverless computing offers an additional opportunity to control the entire development stack, reduce operational costs by efficiently optimizing and managing cloud resources, and provide a platform that encourages the use of additional services in their ecosystem. And reduce the effort required to author and manage cloud-scale applications (Castro et al. 2019).

Based on the serverless computing platform provided by AWS, this paper performs high-dimensional matrix cross-multiply and matrix convolution operations and adjusts different variables. Conclusions are drawn by analyzing the impact of these variables on calculation time and cost, respectively. Then consider the expansion direction of serverless computing based on the conclusion (Villamizar et al. 2017).

2 The Overview of Research Framework

The paper is based on the AWS Lambda function computing service provided by AWS and uses S3 as a storage tool (Sbarski 2017; Anh 2015). For high-dimensional matrix cross-product, suppose that both matrix A and matrix B are N*N matrices, and calculate the result of A*B. By changing the sizes of matrix A and B and the number of worker nodes, observe the time-consuming to complete the entire computing process and the relationship between the communication time and execution time in the worker nodes. For the matrix convolution operation, suppose that matrix A is an N*N matrix and keep the convolution kernel and step size unchanged. By modifying the memory parameters, we can observe the change in running time and try to estimate the operation cost (Lee et al. 2017).

In addition to the above main indicators, observation of indicators such as system stability and worker node start-up delay is required. Some system design factors of AWS Lambda may affect the time when the node starts to work, such as task re-entry due to execution failure, waiting or even serialization due to limited resources, and whether to reuse the previous instance for the warm start, etc.

As shown in Fig. 1, The general process of this work is: (1) the trigger node is responsible for distributing tasks and concurrently scheduling worker nodes to start executing tasks. (2) After the worker node is triggered, download the matrix to be calculated from S3 according to the parameters passed by the trigger. (3) After the worker node completes the calculation, upload the result to S3. (4) At the same time as the calculation, the worker node will also record the time-consuming key links in the execution process and upload them to S3 after the calculation is completed. (5) After the trigger node distributes the task, it reads all the calculation results on S3 by waiting and polling and splices it into a complete result matrix. Since the trigger and worker nodes are asynchronous, the process of uploading monitoring data by worker nodes does not block the entire computing process. (6) The data analysis interface will read the monitoring data uploaded by the worker nodes on S3 and calculate and analyze it to obtain statistical results.

Regarding data chunking design, the matrices are constructed locally and then uploaded to S3. For high-dimensional matrix cross-product, matrix A and matrix B are split by row and column, respectively, and each node computes the result of one of the blocks. For the convolution operation, the result is first blocked and then reversed by the block rules of the original matrix.

Regarding the merge logic, two options are considered: (1) the first solution is that the trigger node waits for some time after distributing tasks and then polls the calculation results of each worker node on S3 according to the index order. If a 404 error occurs, the file does not exist. The trigger node will wait in place until the result is read and the local stitching is completed. And so on until the calculation results of all nodes are read, and the final matrix is obtained. (2) The second solution is that the trigger node waits for some time after distributing the task and then polls the calculation results of each worker node on S3 according to the index order. If a 404 error occurs, the file does not exist. The trigger node maintains a list, records

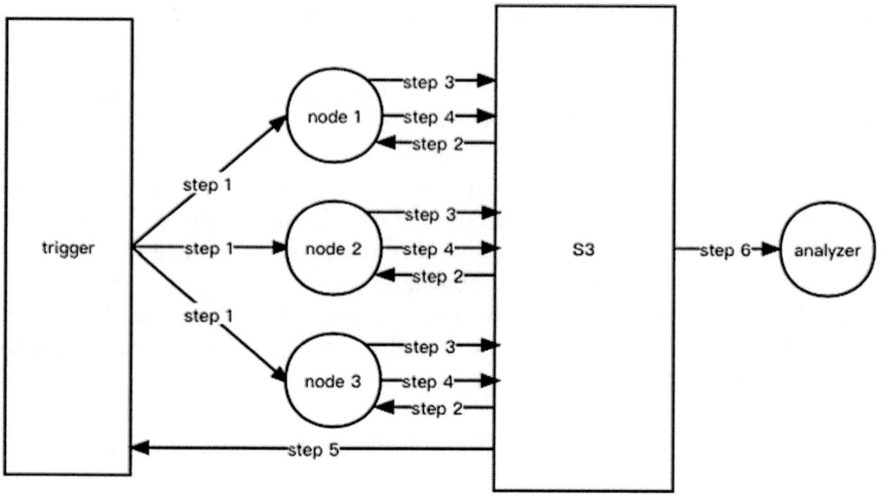

Fig. 1 Framework design

the uncalculated node, and asks for the next node. Nodes that have not completed computations are recorded in the list until a poll is completed. Then the trigger will wait and poll the nodes in the list until it gets all nodes' calculation results and the final matrix.

These two schemes have advantages and disadvantages regarding time and memory usage. In scheme A, the result matrix is spliced in order, and the memory utilization efficiency is high; but the waiting logic is not flexible enough. In the worst case, the first node may Calculate the slowest node, and then the splicing time is $O(N) = \max(\text{Node\_}i) + (N - 1) * \text{merge\_cost}$. The advantage of scheme B is that the waiting logic is more flexible, but because additional maintenance lists are required, and the matrices may not be spliced in order, a large-scale matrix needs to be maintained at the beginning, and the memory usage is relatively high. Considering that the worst-case scenario in Scheme A is extreme, the memory usage problem in Scheme B is inevitable, so Scheme A is adopted as the final merging logic.

3 Results and Analysis

In the first experiment, the number of worker nodes is kept at 100 and the experiment observes the relationships between the matrix size and the time results.

As shown in Table 1, when N = 3000, the time-consuming communication ratio is relatively high, which becomes the main bottleneck of efficiency improvement. As N increases, the proportion of communication time-consuming decreases significantly. When N = 7000, the proportion of communication time-consuming is only 2.1%. Currently, the main bottleneck affecting efficiency is the computing task itself. Due to the design of this experiment, the communication times of each worker node are fixed, so as the data size increases, the impact of communication costs on serverless computing gradually decreases.

Besides, the average is slightly larger than the TP90, indicating that a small amount of long-tailed data takes a long time. In an ideal state, each worker node executes concurrently, and the calculation results of all worker nodes are merged at the end. According to the barrel effect, the most time-consuming node will affect the total time-consuming of the entire work process.

In the second experiment, the matrix size is kept, and the experiment observes the relationships between the number of worker nodes and the time results.

As shown in Table 2, when the number of worker nodes increases from 1 to 9, the experimental results show that the completion time of the entire computing process is significantly shortened, and the time consumption of each worker node also decreases.

When the number of concurrencies is increased from 9 to 16 (AWS default concurrency is 10), since the amount of tasks allocated to the worker nodes is reduced, the time-consuming of the worker nodes is reduced. Still, due to the limited concurrent resources, there is a situation of queuing and waiting. The maximum start-up delay reaches 17 s, which ultimately affects the completion time of the entire calculation process (Gupta et al. 2018).

Table 1 Time cost of multiplications with different matrix sizes

Matrix dimension	3_000	5_000	7_000
Number of worker nodes	100	100	100
Average time spent on worker nodes (s)	1.7359	9.8567	26.9314
The proportion of average communication time of worker nodes (%)	20.6569	5.6567	2.7571
Worker node time TP90(s)	1.5540	9.3481	26.4145
Worker node communication time ratio TP90 (%)	15.4745	3.7998	2.1048
Maximum time spent on worker nodes (s)	2.3791	11.2020	27.8542
The maximum communication time ratio of the worker nodes (%)	37.4940	12.7009	4.3909

Table 2 Time cost of multiplications with different numbers of worker nodes

Matrix dimension	3_000	3_000	3_000	3_000
Number of worker nodes	1	4	9	16
Average time spent on worker nodes	206.2691	53.8583	23.4947	13.7865
Maximum time spent on worker nodes	206.2691	55.8976	23.8825	14.4655
Maximum start-up delay of worker nodes	0	0.1643	0.3953	17.2341
Completion time (s)	206.2691	59.8212	27.8584	32.2963

In the third experiment, the matrix parameters are kept, and the experiment observes the overall execution time and cost estimates by modifying the memory parameters.

According to Lambda's configuration rules, increasing the memory parameter will also increase the CPU resources allocated to the function. As shown in Table 3, when the matrix parameters remain unchanged, increasing the memory parameters can reduce the running time of each node and the time-consuming of the entire process.

At the same time, Lambda's billing rule is to calculate the product of function time and allocated memory. Therefore, the economic cost should also be considered while increasing the memory parameters to shorten the running time. When the memory parameter is increased from 512 to 2048 M, the cost also increases, although the running time is significantly reduced.

Given the small number of matrices used in this experiment, there is no particular significance in monitoring the individual data flows of each node when assigning load, but if large data sets involving high-dimensional matrices are involved, then monitoring the impact on system performance when resource capacity is added or removed respectively would largely avoidable. In addition, due to the limitation on the number of concurrent nodes in this experiment, the resilience of the system when all nodes are allowed to be concurrent can hardly be monitored. The same situation occurs in monitoring the impact of changes in scaling speed and system performance on concurrent workloads because, in an experiment where the number of nodes is the dependent variable, not all nodes are working simultaneously. Contemporary work is not executed together, meaning it is impossible to measure the full system. The fact that contemporary work is not performed together means it is impossible to measure the complete system's scaling efficiency at runtime. However, this monitoring metric can be implemented when more concurrent approvals are requested, which is still possible (Kuhlenkamp et al. 2014; Bermbach et al. 2017).

To control the run time within the specified time (15 min), the experiment sets a computation time monitoring function suitable for this experiment, which also implies that the computation time may increase significantly for datasets of the same latitude but with a larger amount of data. This paper designs a fixed segmentation

Table 3 Time cost and budget of multiplications with different memory parameters

Matrix dimension	1000	1000	1000
Convolution kernel size	5	5	5
Number of worker nodes	4	4	4
Submatrix dimension	502	502	502
Memory parameter setting (MB)	256	512	2048
Average time spent on worker nodes (s)	25.0452	11.9384	3.4543
Completion time (s)	28.0213	14.9330	5.8762
Estimated cost per node (MB*s)	6411.5712	6112.4608	7074.4064

function for the cut phase, as increasing the number of units per cut may significantly increase the run time. If validation is required, a function to monitor the increase in the number of units per cut versus time and a function to compare the difference in run time between nodes was designed.

4 Discussion

According to the experiments, the following questions about serverless computing deserve further discussion.

Extreme cases. Whether worker nodes execute synchronously or asynchronously, the slowest node is one of the bottlenecks in the efficiency and consumption of the system. In large-scale computing, there could be tens of millions of worker nodes, so even a slight possibility will lead to many problem occurrences. Such a situation will increase the completion time of the system and cost more money. For this issue, the solution is that limited computation errors should be allowed, and compensation will be added to the result to guarantee the accuracy of the whole computation and control the budget. Specifically, this solution could be taken in two ways: (1) discard the slow worker nodes, considering they contribute nothing to the result. (2) Construct a result from other worker nodes and regard it as the contribution from the slow worker nodes.

Imagine that there is an application with tens of millions of users in which the server will profile and analyze users based on their preferences for further product optimization. The whole process could be summarized as follows: (1) the server, the master node, sends contents to users' phones which are worker nodes. (2) These worker nodes calculate results about users' preferences according to their choices on these contents and return the data to the master node. (3) The master node receives data from worker nodes and calculates the result to get the integrated portrait of users.

If several worker nodes execute too slowly, two methods are available: (1) when the computation resource and time reach the upper limits, the master could stop waiting for the nodes still calculate, and their contributions will not be considered. The master node will work out the result with the rest of the nodes. In more extreme cases, when the cost is more valued than the accuracy, the system could stop waiting when the proportion of finished nodes reaches a particular threshold. In practical production, if resource and time limits, the proportion of finished nodes could be adjusted dynamically, multiple small follow tests are accessible to find the optimal dynamical scheme. On the other hand, the disadvantage of this scheme is that if the nodes discarded are not independent and identically distributed, systematic bias will occur. For instance, if users in an area constantly fail to finish the computation due to weak network conditions, the result will miss the contribution of users from this area leading to a bias in the portrait. For this error, if the master node could cache the temporary results of slow nodes and set them as the initial status of the next round of computation, such slow nodes could catch up with the whole procedure. However, the feasibility of this solution depends on the algorithms on worker nodes and could be considered in certain situations. (2) When the computation resource and time reach the upper limits for slow nodes, the master node can stop waiting and calculate a relative representative result from the rest of the nodes as the results of the slow nodes. Thus, the slow nodes will equally join in the final integration. Apart from the former, worker nodes are considered the most in this scheme. But in some scenarios, the result is homogeneous because the results of the finished nodes are strengthened, and the actual results of the slow nodes could be the opposite. Nevertheless, this problem is avoidable in practice. Since technology products update cyclically, the result of the last version could be the substitute result. Fully verified in the last version, this choice is more stable and safer though it will slow down the updates of the product.

Queuing problems. Resource Limits like physical machines breaking down and traffic surges like promotion strategies attracting new clients are unavoidable in practice. Regardless of these emergencies, the resource provided by manufacturers is normally limited. Even if the workload exceeds the upper limits of the system, which is regarded as a problem in experimental circumstances, such situations seldom cause exceptions in practice. In parallel computing, the time cost of each worker node varies due to different computing conditions. The system could re-distribute tasks to worker nodes with less workload and alleviate queuing problems. Besides, background computing, animation, and other product strategies could conceal the delay caused by queuing from users.

Monitoring indicators. Because the core computation is executed on worker nodes, monitoring this process is indispensable for researchers to handle and optimize the system. Based on the experiment, the following indicators are essential for a single worker node: (1) the execution time of each worker node: This is a fundamental indicator to demonstrate the workload of each worker node. (2) The proportion of the communication time to the execution time of each worker node: Communication between worker nodes and the master node will bring additional I/O costs. So serverless computing is more efficient when the proportion of the communication time decreases. (3) The economic cost of the computation of each worker node: the charging rules depend on the manufacturers.

What's more, the following indicators are required for all the worker nodes as a whole: (1) the largest time cost of the start-up delay on worker nodes: start-up delay caused by exceptions such as cold start-up, retry and queuing will postpone the finish time of the execution on the single worker node. Hence the largest delay harms the total computation time. (2) The failure rate of the execution on worker nodes: Whether the execution will fail or not depends on not only the stability of the service provided by the manufacturers but also the specific algorithm on the worker node. Therefore, this indicator is necessary for a serverless computing design. (3) The total time cost of the computation: Because of the resource limits, the trend of the total computation time could be opposite to the trend of the computation time on each worker node, which is indicated in the experiment.

5 Conclusion

The experiment discusses before attempts to conduct serverless computing with different parameters and theoretically analyzes the results. For large-scale applications, serverless computing is more efficient, with more workload in limited situations. When the resource reaches the upper limits, or the computing circumstances change, the performance of serverless computing could be worse than expected. Higher efficiency demands more resources. Moreover, the experiment indicates that economic cost and computing stability are critical issues for the further expansion of serverless computing. Based on practice in the IT industry, solutions for fundamental problems about efficiency, resource, and monitoring in serverless computing are proposed. The experiment should be enriched for future work to simulate real industrial situations as possible and reveal more logic behind practical exceptions. Furthermore, the above schemes could be implemented on such comprehensive experimental configurations to verify their effectiveness.

References

Anh TN (2015) Amazon S3 simple storage service—guide

Baldini I, Castro P, Chang K et al (2017) Serverless computing: current trends and open problems. In: Research advances in cloud computing. Springer, Singapore, pp 1–20

Bermbach D, Wittern E, Tai S (2017) Cloud service benchmarking. Springer International Publishing, New York, NY

Castro P, Ishakian V, Muthusamy V et al (2019) The rise of serverless computing. Commun ACM 62(12):44–54

Gupta V et al (2018) Oversketch: approximate matrix multiplication for the cloud. In: 2018 IEEE international conference on big data (Big Data). IEEE

Kuhlenkamp J, Klems M, Röss O (2014) Benchmarking scalability and elasticity of distributed database systems. Proc VLDB Endowment 7(12):1219–1230

Lee K, Suh C, Ramchandran K (2017) High-dimensional coded matrix multiplication. In: 2017 IEEE international symposium on information theory (ISIT). IEEE, pp 2418–2422

Sbarski P (2017) Serverless architectures on AWS: with examples using AWS Lambda

Villamizar M, Garcés O, Ochoa L et al (2017) Cost comparison of running web applications in the cloud using monolithic, microservice, and AWS Lambda architectures. SOCA 11(2):233–247

Werner S, Kuhlenkamp J, Klems M, Müller J, Tai S (2018) Serverless big data processing using matrix multiplication as example. In: 2018 IEEE international conference on big data (Big Data)

Design and Application of Data Management System for the Coronavirus Pandemic

Peicheng Yao

Abstract COVID-19 has quickly become a global health emergency ever since it was first discovered in 2019. The increasingly large amount of data emerging from the pandemic has created a request for effective database management systems for tracking and managing COVID-19 related information. This paper introduces a SQL-based database management system that tracks the important information related to COVID-19, with a focus on managing data related to people and their activities. The system is designed by first translating a composed list of mission objectives and requirements into an Entity-Relationship Diagram (ERD), which establishes and visualizes the structure of the system. From the finalized ERD, data tables are created for each of the entities in the ERD, which are then created inside an Oracle SQL database, and actual data entries are also populated into the database using SQL statements. After the creation of tables and insertion of data entries, the database is set up and ready for queries. In order to make the system operable for non-SQL experts, a graphical interface is developed using Python and PyQT5. The graphical interface includes a login window, a main window for selecting entities, and operation windows for each of the entities in the database. Inside each operation window, users can view/add to/remove from/modify data entries in the database. Ultimately, the goal of this work is to provide a robust and efficient database management system that authorities can utilize to track, analyze and manage the large amount of data that emerged from the COVID-19 pandemic.

1 Introduction

First identified in December 2019 in Wuhan, China, the novel coronavirus (COVID-19) has quickly spread across the globe as one of the largest pandemics the human race has faced in recent decades. As part of the coronavirus family, COVD-19 is an RNA virus that, once infected, causes illness in the respiratory system along with

P. Yao (✉)
Department of Computer Science, University of British Columbia, Vancouver, BC, Canada
e-mail: yaop@students.cs.ubc.ca

S. Yadav et al. (eds.), *Proceedings of 2nd International Conference on Artificial Intelligence, Robotics, and Communication*, Lecture Notes in Electrical Engineering 1063, https://doi.org/10.1007/978-981-99-4554-2_10

fever, dry cough, dyspnea, and other more severe symptoms (Yang et al. 2020a). Since it is so highly contagious and causes severe illnesses, COVID-19 has immediately developed into a global public health emergency since its first discovery. According to BBC News, on January 23rd, 2020, only 2 days before the Chinese New Year, Wuhan went completely silent with 11 million people put under quarantine. Until June, the city was completely sealed off from the rest of the world (BBC 2021). According to the statistics provided by the World Health Organization (WHO), there are over 600 million confirmed cases of COVID-19 and over 6.4 million deaths reported across the globe as of September 4th, 2022 (World Health Organization 2022). The COVID-19 pandemic plagued the world and posed a threat to the entire human race. As a result, the aim to reduce the spread and cure the infected has become a shared value across all nations. In order to control the spread of the virus and help society to recover from the crisis, researchers from all over the world are seeking to collect and analyze this enormous amount of data generated by the outbreak, which includes who is infected, who has contact with infected people, where the infected people have been to, which cities need to be locked down, etc. With this increasingly large volume of data emerging from the pandemic, a robust data storage system is called for. More specifically, a Database Management System (DBMS) that is efficient, secure and simple to use is needed for the authorities to utilize to track and analyze the vast amount of data coming from the pandemic.

A DBMS is a platform for clients to use to access and interact with databases where data are stored. The main two types of DBMS are relational databases with Structured Query Language (SQL) and non-relational databases, namely NoSQL databases. Research have shown that SQL databases are better at dealing with structured data and joint queries, while NoSQL databases are better with unstructured data with no need to join queries (Antas et al. 2022). SQL databases are also found to be easier to use, execute and maintain, and more secure (Ali et al. 2019). As a result, this study primarily focuses on the applications of SQL databases instead of NoSQL databases. Since the outbreak in 2019, researchers have conducted studies on how to make effective use of DBMS in the context of COVID-19 to better store, analyze and manage the related data. In 2020, Eman Khashan et al. proposed a storage framework (COVID-QF) that supports both SQL and NoSQL databases to process complex queries with COVID-19 datasets in an attempt to help with the problems generated by the diversity and volume of COVID-19 related data (Khashan et al. 2020). COVID-QF is a well-designed system that integrates SQL and NoSQL together to utilize both databases' advantages in the pursuit of finding the most effective system for dealing with COVID-19 datasets. Antibody therapeutics is also a vital aspect of controlling the spread of the virus. In 2020, the Chinese Antibody Society initiated the "Tracker" program, which is a global online database of antibody therapeutics. It made use of SQL databases and the system was developed to help researchers analyze and demonstrate COVID-19 antibody developments (Yang et al. 2020b). Similar efforts can also be seen in Dr. Abhishek Thote and Dr. Rajesh Patil's study on the development of a SQL-based DBMS that stores COVID-19 related information such as hospitals, police departments, travel agencies, patients, and others (Thote and Patil 2022). This system can be used by authorities to store, track and analyze data on people's conditions under the pandemic and react accordingly. Many more examples can be found across the globe where researchers are working with dedication to utilize and apply the abilities of DBMS in the COVID-19 situation. However, most of the proposed systems appear to be either too shallow in depth or too limited in scope, failing to capture all of the significant aspects of data in the COVID-19 context. This provided motivation for a comprehensive high-level DBMS that captures most, if not all, of the major pieces of information a governing body should be interested in when trying to put the pandemic under control.

Consequently, this paper proposes a SQL-based DBMS for tracking all the important facets of information related to COVID-19, with a focus on the people and their activities. This system is developed to provide effective support for the authorities to access, monitor, and analyze the data of interest in the COVID-19 pandemic, and ultimately to help them make progress in tackling the issue. This system is first designed by generating a list of mission objectives and data requirements that this system aims to accomplish. Then, the requirements are visualized through an Entity-Relationship Diagram (ERD), which is essentially a diagram laying out all the entities of interest as well as their relationships with one another. This includes the people, the cities, the medical institutions, and so on. When the ERD is finalized, it is transformed into a set of data tables that functions as our primary model for the data. The tables capture all the information needed from the ERD to add the data structure into a SQL database. The relations are partitioned into tables together with constraints that describe relationships between them. After that, the tables are added to an Oracle SQL database using SQL statements and are connected to each other through SQL constraints. The database is where all the operations on the data take place, including populating data entries into the database, modifying data, fetching data of a particular range, and so on. After the database is set up, a graphical user interface (GUI) is developed for clients to use to operate the database. This DBMS keeps track of most information of interest related to COVID-19, including the condition of every person, the risk level of each city, the vaccinations, quarantine information, travel records, hospitalizations, and more. The database also records information about polymerase chain reaction (PCR) testing, which is the mainstream virus testing method for COVID-19, and is considered as the "gold standard" of virus testing due to its rapid detection, high sensitivity, and specificity (Tahamtan and Ardebili 2020). The goal of this study is to create a database that provides functionalities that are vital but absent in past works and ultimately support authorities to keep track of important data of the COVID-19 pandemic.

The rest of the paper is organized as follows: In Sect. 2, the paper goes through the design process of the system including the creation of the system ERD and data tables. Section 3 demonstrates the detailed implementation of the system and presents the SQL statements for building the database as well as the GUI for user interaction. Finally, Sect. 4 concludes the paper with a conclusion and discussion for future work.

2 Design of the COVID-19 Data Management System

A. System Mission Objectives and Requirements

The primary objective of this system is to maintain a comprehensive set of data on the people within a region, the cities and the medical institutions. Capturing the relations among the three will provide an information network that can be queried easily to track down any specific piece of data. The major mission objectives are listed as follows:

- To maintain information on people, including the condition of everyone (infected, safe at risk, cured, dead), the quarantine information, the vaccination, travel records, current and home cities, insurance companies, and so on.
- To maintain information on cities, including the risk level of each city, the number of total confirmed/current confirmed/cured cases/deaths, and the medical institutions in each city.
- To maintain information on medical institutions, including PCR test centers, hospitals for vaccinations and hospitalizations, and medical labs for processing PCR tests and producing vaccines. The system also needs to keep track of every appointment with either a PCR test center or a hospital.

B. Overview of System Structure

From the mission objectives, the system design can be abstracted into three main modules of focus, namely People, City and Medical Institution. The three modules are further divided into more detailed sub-modules of functionality, each covering a specific mission objective described in the previous section. As shown in Fig. 1, People's Information.

Management consists of functionalities such as Current/Home City Tracking, Travel Record Management, Vaccination Record Management, Quarantine Management, and Condition Management. The City Information Management module is divided into Local Medical Institution Tracking, Case Count Management, and Risk Level Management. Finally, the Medical Institution Information Management module includes Medical Lab Management, Hospital Management, and PCR Test Center Management.

C. Visualization of Entities and Relationships with ERD

In order to visualize the mission objectives and design requirements, an ERD is developed with all the objects of interest and their relationships with each other. In database developments, the approach of modeling the system with an ERD is widely used because it is easy to understand, strong at modeling real-world problems, and can be readily translated into a database schema. Essentially, an ERD models the real-world context into a set of business entities, a collection of relationships in between them, and attributes that describe each of the entities (Song et al. 1995). By providing an interconnected mapping of the objects in the system design, the ERD serves the function of a general overview of the project. As can be seen in Fig. 2, the ERD records

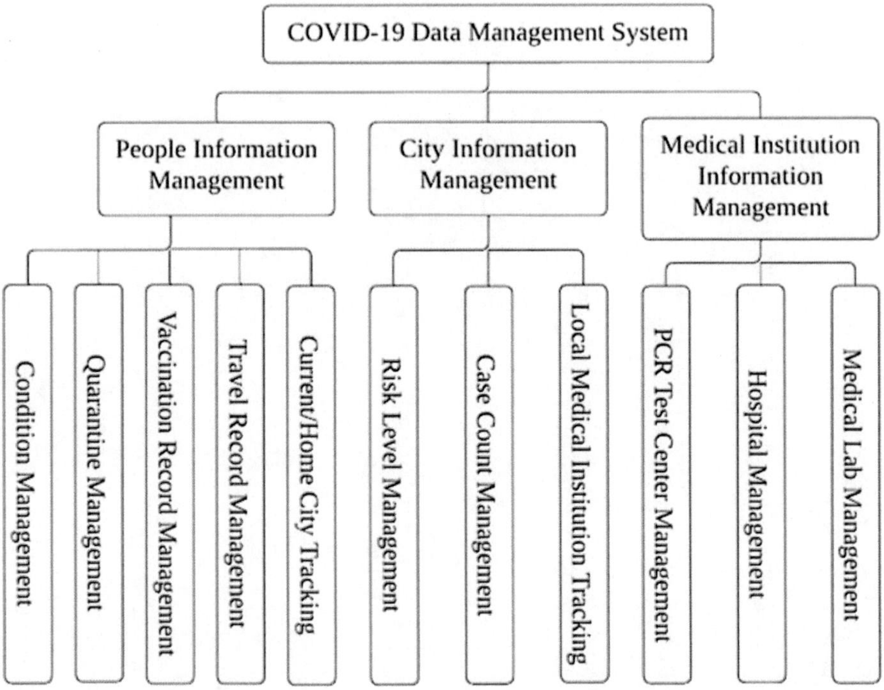

Fig. 1 Overview of system structure

all the entities of interest within the scope of the system design. In the right center of the ERD is the Person entity, consisting of all the important attributes about a person such as a person's personal information, condition, vaccination information, contact number, whether this person needs quarantine, and whether he/she is a traveler from another city. The Person entity is connected to other entities by its relationships with them respectively: people use insurance companies, have quarantine information if needed, and have travel records that document a person's travel paths within the past 14 days, etc.

Connected by the Travel Record entity with the People entity is the City entity, which includes the name, population, risk level and the case counts of a city. Cities are also recorded in the Travel Record entity together with the People entity, which forms a record table with all the travel activities of everyone within the past 14 days. A person's current location and original city are also recorded through the connections between People and City entities.

The entity group of medical institutions can be found to the left of the diagram including the PCR Test Center entity, the Medical Lab entity, and the Hospital entity. The ERD keeps track of the institutions' basic information along with their capabilities and capacities. The Medical Lab entity is connected to both PCR Test Center and Hospital because medical labs do the job of both processing tests for PCR centers

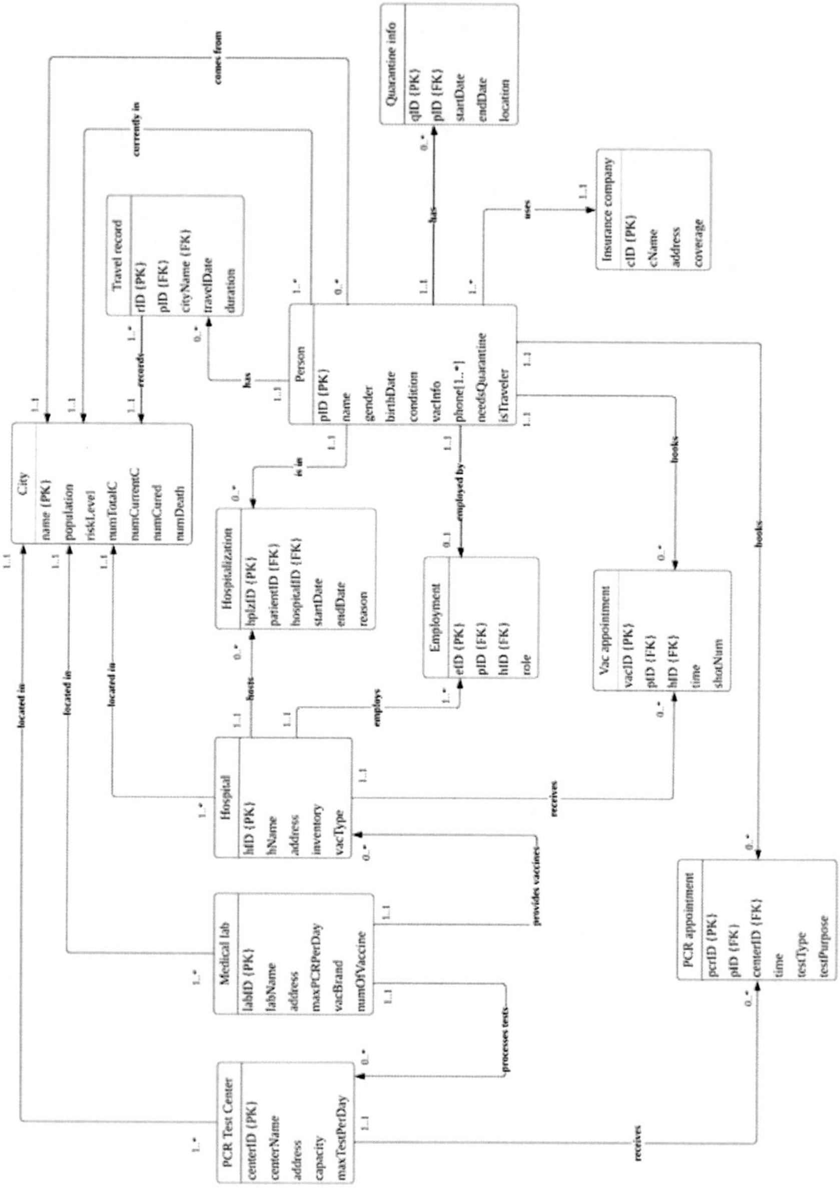

Fig. 2 ERD of the system

and producing vaccines for hospitals. The connections going out of the medical institutions describe the relationships between the institutions and the other entities. For instance, people can book appointments for PCR tests and vaccines with PCR test centers and hospitals, respectively. This information is stored in the PCR Appointment and Vac Appointment entities with the appointment time and other related information. People can also work for hospitals and the employment information is recorded by the Employment entity that connects the People and Hospital entities. Last but not least, the hospitalization information is also recorded in the Hospitalization entity which consists of the information of the patient and the hospital, the start and end dates of the stay, as well as the reason for the hospitalization.

D. Design of Data Tables of the SQL Database

After the finalization of the ERD design, the structure of the system is ready to be created in the SQL database. The process of developing a SQL database from the ERD requires a transformation from the ERD into a set of data tables that records all the attribute information for each entity. This table set can be later added to the database using SQL statements. Shown below are some of the major tables in the table set.

(1) *People Data Table*

Table 1 shows the table containing information about the People entity, consisting of all the attributes seen in the ERD of People with their designated data types in the SQL database. The type of "VARCHAR2" here basically means a text entry for those not in the know. The "Allowed Null" column indicates whether or not this attribute can be an empty value in the database. A description of the attribute is also included in the table.

(2) *City Data Table*

Table 1 People data table

Attribute name	Data type	Allowed null	Description
pid	NUMBER	NOT NULL	Personal ID
name	VARCHAR2(30)	NOT NULL	Person's name
gender	VARCHAR2(1)	NOT NULL	Person's gender
birthDate	DATE	NOT NULL	Person's date of birth
condition	VARCHAR2(30)	NOT NULL	Person's condition
vacInfo	VARCHAR2(30)	NULL	Person's vaccination information
needsQuarantine	NUMBER(1)	NOT NULL	Whether or not this person needs to be quarantined
isTraveler	NUMBER(1)	NOT NULL	Whether or not this person is a traveler
currentCity	VARCHAR2(30)	NOT NULL	Person's current city
originCity	VARCHAR2(30)	NOT NULL	Person's city of origin
insuranceCo	NUMBER	NOT NULL	Person's insurance company ID

Table 2 displays the City table of the database with the name, population, and counts of different types of COVID-19 cases.

(3) *PCR Test Center Data Table*

Table 3 is the table for the PCR Test Center entity, derived from the attributes in the ERD. As can be seen, the cityName attribute represents a relationship between the PCR Test Center and the City entities: the information about the location of a PCR center is stored as the city name inside the PCR Test Center table. A similar representation of a relationship with the Medical Lab entity can be seen in the labID attribute as well.

(4) *PCR Appointment Data Table*

Table 2 City data table

Attribute name	Data type	Allowed null	Description
name	VARCHAR2(30)	NOT NULL	City name
population	NUMBER	NOT NULL	City population
riskLevel	VARCHAR2(30)	NOT NULL	City risk level
numTotalC	NUMBER	NOT NULL	Number of total confirmed cases in the city
numCurrentC	NUMBER	NOT NULL	Number of current confirmed cases in the city
numCured	NUMBER	NOT NULL	Number of cured cases in the city
numDeath	NUMBER	NOT NULL	Number of total deaths in the city

Table 3 PCR test center data table

Attribute name	Data type	Allowed null	Description
centerID	NUMBER	NOT NULL	PCR center ID
centerName	VARCHAR2(30)	NOT NULL	PCR center name
address	VARCHAR2(50)	NOT NULL	PCR center address
capacity	NUMBER	NOT NULL	PCR center capacity
maxTestPerDay	NUMBER	NOT NULL	Number of maximum number of tests that can be performed by this center
cityName	VARCHAR2(30)	NOT NULL	Name of the city this center is located in
labID	NUMBER	NOT NULL	ID of the lab that processes this center's PCR tests

Table 4 PCR appointment data table

Attribute name	Data type	Allowed null	Description
pcrID	NUMBER	NOT NULL	Appointment ID
pid	NUMBER	NOT NULL	ID of the patient
centerID	NUMBER	NOT NULL	ID of the PCR center
time	DATE	NOT NULL	Date and time of the appointment
testType	VARCHAR2(30)	NULL	Type of the PCR test (individual/group)
testPurpose	VARCHAR2(50)	NULL	Purpose of the PCR test

Table 4 shows the table of the PCR Appointment table with the related attributes including the IDs of the appointment, the patient, the test center, as well as the appointment time, the type of the test, and the purpose of the test.

3 Implementation of the COVID-19 Data Management System

In the implementation of the system, data tables designed in the previous section are created in an Oracle SQL database using SQL statements, after which the actual data entries are populated into the database. From the database, data can be retrieved by performing SQL queries with specific focuses and constraints. To make the system operable for non-SQL engineers, a GUI is also developed using Python to help users interact with the database.

A. SQL Database Setup

Tables can be created in the SQL database using SQL's CREATE TABLE statement with entries from the data tables in the previous section as inputs. After all the tables are created successfully, the system's structure is set up in the database. Figure 3 shows the CREATE TABLE command for the Person entity including the attributes to be created for this table and CONSTRAINT keywords that indicate the primary key and foreign keys of the table. These SQL constraints establish the connections between each table. In this case, the primary key is the ID of the person and the foreign keys are the current city attribute, the origin city attribute, and the insurance company ID attribute, referring to the City and Insurance Company tables respectively.

Similarly, as shown in Fig. 4, the CREATE TABLE command for City consists of attributes identical to those in the ERD, together with the primary key constraint, which is the name of the city. It is also notable that the City table is referred to in the Person table, but not vice versa. This is because the relationship between cities and people is tracked on the people's side, meaning that it is which city a person is in that is of interest, not the other way around.

After all the tables are created in the database, data entries are added into the database using SQL's INSERT INTO … VALUES command. This command inserts

```
CREATE TABLE Person (
    pID number NOT NULL,
    name varchar2(30) NOT NULL,
    gender varchar2(1) NOT NULL,
    birthDate date NOT NULL,
    condition varchar2(30) NOT NULL,
    vacInfo varchar2(30),
    needsQuarantine number(1) NOT NULL,
    isTraveler number(1) NOT NULL,
    currentCity varchar2(30) NOT NULL,
    originCity varchar2(30) NOT NULL,
    insuranceCo number NOT NULL,
    CONSTRAINT person_pk PRIMARY KEY (pID),
    CONSTRAINT person_fk_1 FOREIGN KEY (currentCity)
                          references City(name) ON DELETE CASCADE,
    CONSTRAINT person_fk_2 FOREIGN KEY (originCity)
                          references City(name) ON DELETE CASCADE,
    CONSTRAINT person_fk_3 FOREIGN KEY (insuranceCo)
                          references InsuranceCompany(cID) ON DELETE CASCADE
);
```

Fig. 3 Person entity CREATE TABLE command

Fig. 4 City entity CREATE TABLE ccommand

```
CREATE TABLE City (
    name varchar2(30) NOT NULL,
    population number NOT NULL,
    riskLevel varchar2(30) NOT NULL,
    numTotalC number NOT NULL,
    numCurrentC number NOT NULL,
    numCured number NOT NULL,
    numDeath number NOT NULL,
    CONSTRAINT city_pk PRIMARY KEY (name)
);
```

data entries with the input parameters into the created tables in the database and these data can be operated upon by queries. Figure 5 shows a list of INSERT INTO … VALUES commands for the Person table. The input parameters follow the same order as specified in the CREATE TABLE command for this table. In this snapshot, 10 entries are populated into the Person table with different parameters.

The same pattern can be found in the commands for the City entity in Fig. 6. One thing to note is that the parameter type for each attribute must be coherent with the specification in the CREATE TABLE command.

B. SQL Queries for Data Retrieval

After the creation of the tables and insertion of the data entries, the SQL database is set up and ready for queries. Queries are SQL statements that select information based on constraints from the database. SQL database support complex user queries that help users retrieve data of interest with no excess information. A simple SQL query can be found in Fig. 7, which returns the list of names of all the travelers

```sql
INSERT INTO Person VALUES(1, 'MICHAEL', 'M', TO_DATE('1996/04/03', 'yyyy/mm/dd'),
                         'SAFE', '', 0, 0, 'NEW YORK', 'NEW YORK', 5);
INSERT INTO Person VALUES(2, 'JACK', 'M', TO_DATE('1999/06/23', 'yyyy/mm/dd'),
                         'AT RISK', 'VAC1', 1, 1, 'NEW YORK', 'MIAMI', 2);
INSERT INTO Person VALUES(3, 'AMY', 'F', TO_DATE('2002/11/05', 'yyyy/mm/dd'),
                         'SAFE', 'VAC1', 0, 0, 'LOS ANGELES', 'LOS ANGELES', 2);
INSERT INTO Person VALUES(4, 'EMMA', 'N', TO_DATE('1987/12/30', 'yyyy/mm/dd'),
                         'SAFE', 'VAC2', 0, 0, 'LAS VEGAS', 'LAS VEGAS', 10);
INSERT INTO Person VALUES(5, 'JASON', 'M', TO_DATE('2000/11/21', 'yyyy/mm/dd'),
                         'CURED', 'VAC1', 0, 1, 'NEW YORK', 'BOSTON', 6);
INSERT INTO Person VALUES(6, 'CHLOE', 'F', TO_DATE('1990/06/06', 'yyyy/mm/dd'),
                         'INFECTED', '', 1, 1, 'MIAMI', 'NEW YORK', 2);
INSERT INTO Person VALUES(7, 'ALICE', 'F', TO_DATE('2005/01/05', 'yyyy/mm/dd'),
                         'SAFE', 'VAC2', 0, 1, 'BOSTON', 'NEW YORK', 3);
INSERT INTO Person VALUES(8, 'HANNAH', 'F', TO_DATE('1972/01/20', 'yyyy/mm/dd'),
                         'INFECTED', '', 1, 0, 'CHICAGO', 'CHICAGO', 3);
INSERT INTO Person VALUES(9, 'JOHNSON', 'N', TO_DATE('2001/12/02', 'yyyy/mm/dd'),
                         'SAFE', 'VAC3', 0, 0, 'NEW YORK', 'NEW YORK', 7);
INSERT INTO Person VALUES(10, 'JOHN', 'M', TO_DATE('2000/11/21', 'yyyy/mm/dd'),
                         'DEAD', '', 0, 1, 'SEATTLE', 'MIAMI', 2);
```

Fig. 5 Person entity INSERT INTO…VALUES command

```sql
INSERT INTO City VALUES('NEW YORK', 8000000, 'HIGH', 5000000, 3000000, 1500000, 500000);
INSERT INTO City VALUES('CHICAGO', 3000000, 'MEDIUM', 700000, 300000, 300000, 100000);
INSERT INTO City VALUES('SAN FRANCISCO', 900000, 'MEDIUM', 200000, 100000, 50000, 50000);
INSERT INTO City VALUES('SAN DIEGO', 1400000, 'LOW', 200000, 50000, 140000, 10000);
INSERT INTO City VALUES('BOSTON', 700000, 'LOW', 200000, 30000, 150000, 20000);
INSERT INTO City VALUES('LOS ANGELES', 4000000, 'HIGH', 2000000, 1000000, 600000, 400000);
INSERT INTO City VALUES('SEATTLE', 740000, 'MEDIUM', 300000, 130000, 120000, 50000);
INSERT INTO City VALUES('PHILADELPHIA', 1600000, 'MEDIUM', 600000, 200000, 350000, 50000);
INSERT INTO City VALUES('LAS VEGAS', 640000, 'LOW', 120000, 50000, 60000, 10000);
INSERT INTO City VALUES('MIAMI', 440000, 'HIGH', 280000, 140000, 110000, 30000);
```

Fig. 6 City entity INSERT INTO…VALUES command

currently in New York. The SELECT keyword indicates which attributes need to be returned. The FROM keyword is followed by the table this query is requesting data from. The WHERE keyword identifies any constraints this query has. In this case, the query is selecting all the names of the entries from the Person table with the constraint that the entry's isTraveler attribute field is 1 and currentCity attribute field is "NEW YORK".

Performing this query on the established SQL database produces a result table containing the requested data. Figure 8 is the resulting data table of the query. As can be seen, three names are returned, meaning that in the current dataset of the system, these three people are all travelers currently in New York.

Fig. 7 Travelers in New York SQL query

```sql
SELECT name
FROM Person
WHERE isTraveler = 1 AND currentCity = 'NEW YORK'
```

Fig. 8 Travelers in New York SQL query result

NAME
JACK
JASON
JACKSON

Figure 9 demonstrates a more complex query that finds the latest hospitalization case due to surgery done at hospital No. 3. This query uses the keyword INNER JOIN to join the Hospitalization table and the Hospital table in order to access data from both tables. This is done through the relationship between the two tables and is reflected in the ON statement after INNER JOIN. This query is also a nested query, meaning that it has a subquery inside it, which is often used to find data that outstand. In this case, the subquery after the ALL keyword returns all the end dates of surgery related hospitalization records at hospital No. 3. This query then finds all the surgery related hospitalization records at hospital No. 3 that has an end date later than all the end dates in the result dataset of the subquery. By using keywords like joining tables and nesting queries, this rather complex query successfully finds the surgery related hospitalization record in hospital No. 3 with the latest end date.

Putting this query into the database gives a result table seen in Fig. 10. As shown, there is only one data entry that satisfies the query constraints. As the asterisk in the SELECT statement implies, the result table displays all information of the result data entry.

```
SELECT Hospitalization.*
FROM Hospitalization
INNER JOIN Hospital ON Hospitalization.hospitalID = Hospital.hID
WHERE Hospital.hName = 'HOSPITAL 3' AND Hospitalization.reason LIKE '%SURGERY%'
      AND Hospitalization.endDate >= ALL(
    SELECT endDate
    FROM Hospitalization
    INNER JOIN Hospital ON Hospitalization.hospitalID = Hospital.hID
    WHERE Hospital.hName = 'HOSPITAL 3' AND Hospitalization.reason LIKE '%SURGERY%'
)
```

Fig. 9 Latest hospitalization SQL query

HPLZID	PATIENTID	HOSPITALID	STARTDATE	ENDDATE	REASON
5	11	3	01-JUL-22	20-AUG-22	BRAIN SURGERY

Fig. 10 Latest hospitalization SQL query result

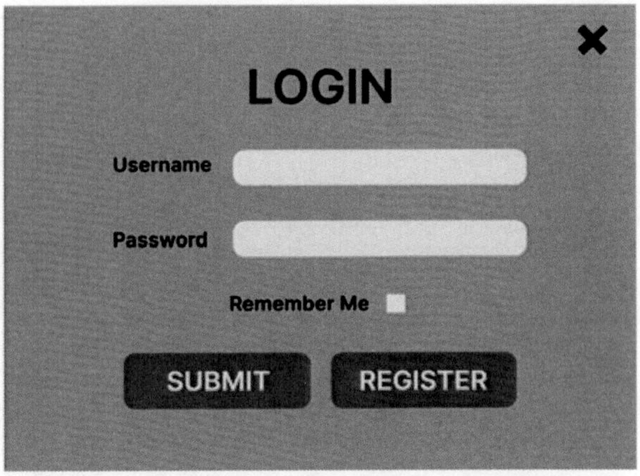

Fig. 11 GUI login window

C. Graphical Interface for the Ease of Use

In order to make the system accessible to users with no prior knowledge of SQL, a graphical interface is developed using Python and PyQT5 plug-in. The GUI includes a login window and an operation window for each of the entities. In each operation window, a user can view/add to/remove from/modify the existing entries of the entity. This change is synchronized with the SQL database through the cx_Oracle module that connects the GUI program to the Oracle SQL database. Figure 11 displays the login window of the program. As can be seen, a user needs to input both a username and a password into the program. The user input is verified in the backend with a database storing all valid login credentials. If the user input matches any existing valid login information in the database, the login is successful.

After a user successfully logs into the program, the user will be directed to the main window of the system as shown in Fig. 12.

As shown, there are 12 buttons on the main window, each directing to an operation window for a specific entity in the system. On this page, a user can navigate back and forth between operation windows to perform actions on different entities such as viewing the existing data entries, adding a new data entry, removing an existing data entry or modify an existing data entry. For instance, clicking on the "View/Add/Remove/Modify Person" button directs the user to the Person entity's operation window shown in Fig. 13. Any added data entries are displayed in the form to the right. Input entries to the left of the window are for collecting parameters when adding new data entries to the database. As can be seen, the three buttons at the bottom of the window indicate the operations that can be performed on the database. Users can remove or modify any existing data entries by selecting the entry in the display form and operating accordingly. It is important to note that because the currentCity, originCity and insuranceCo attributes are references to other tables in the database as discussed before, when adding a new Person entry, these three fields can only take on values that are already in the database. For instance, if "Seattle" is the only City entry currently in the database, then the currentCity and originCity fields of

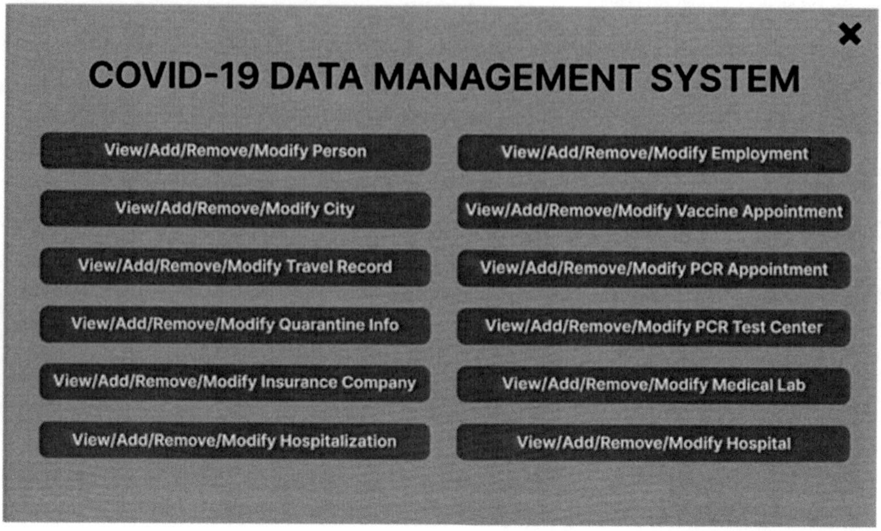

Fig. 12 GUI main window

any Person entry being added can only be "Seattle". This behavior is seen across all tables in the database where references to other tables exist.

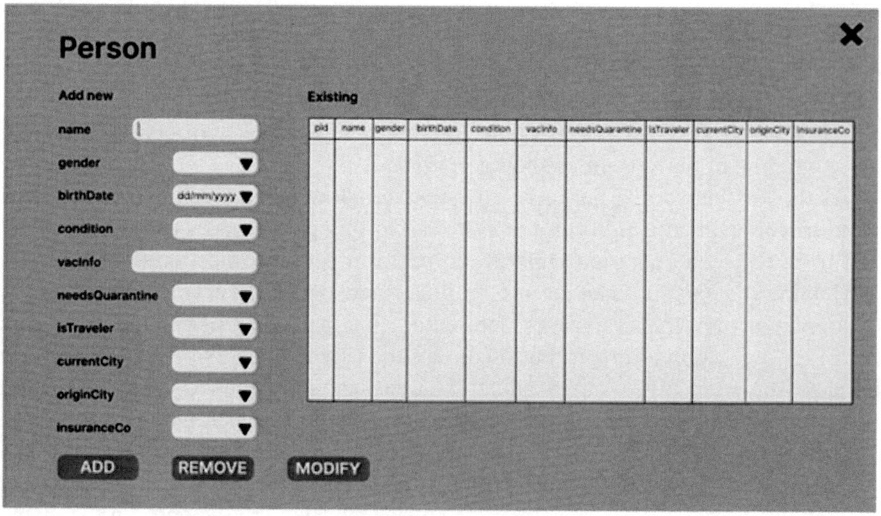

Fig. 13 GUI person entity operation window

4 Conclusion

Taking control of the spread of the virus has been a shared goal across all nations ever since the COVID-19 pandemic initially broke out. The increasingly large volume of data that merged from the crisis had called for new data storage systems to be developed. As a response to this request, this paper introduced a SQL database management system designed for the various types of data related to the COVID-19 pandemic. The database system went through a sophisticated design and development process to ensure a robust outcome. Prior to the design, mission objectives and requirements of the system are developed and altered to make sure they cover all aspects of the proposed system. The structure of the database was then established by sketching out an ERD for the system, which was then converted into data tables that serves the purpose of modeling the entities in the database. After the data tables were finalized, they were created in the SQL database through SQL's CREATE TABLE statements, which were followed by the SQL's INSERT INTO ... VALUES statements that populated actual data entries into the database. At this point, the database is set up and users can perform queries to extract information of interest from the database. However, to make the system useable for users with no knowledge of SQL statements, a graphical interface was also developed that connects the user operation window with the backend database. The GUI was developed with Python and PyQT5, and it made the system more interactable.

Although the proposed system covers a broad range of data related to COVID-19, the system is limited due to the limitations of SQL databases. Compared to NoSQL databases, SQL databases are strong at structured data and complex queries, while NoSQL databases are better at unstructured data and flexible data models. As the volume and diversity of data that emerged from the pandemic continue to grow, a request can be seen for a more versatile DBMS that integrates both SQL and NoSQL databases together to enable optimal effectiveness and efficiency. Future research is expected to focus on utilizing the advantages of both SQL and NoSQL databases to produce systems of maximum capabilities.

References

Ali W, Shafique MU, Majeed MA, Raza A (2019) Comparison between SQL and NoSQL databases and their relationship with Big Data Analytics. Asian Journal of Research in Computer Science 4(2):1–10

Antas J, Silva RR, Bernardino J (2022) Assessment of SQL and NoSQL systems to store and minc COVID-19 data. Computers 11(2)

BBC (2021) Wuhan lockdown: a year of China's fight against the Covid pandemic. BBC News

Khashan EA, Eldesouky AI, Fadel M, Elghamrawy SM (2020) A Big Data based framework for executing complex query over COVID-19 datasets (COVID-QF). arXiv

Song I, Evans M, Park EK (1995) A comparative analysis of Entity-Relationship Disgrams. J Comput Softw Eng 3(4):427–459

Tahamtan A, Ardebili A (2020) Real-time RT-PCR in COVID-19 detection: issues affecting the results. Expert Rev Mol Diagn 20(5):453–454

Thote AM, Patil RV (2022) Concept structure of database management system (DBMS) portal for real-time tracking and controlling the spread of coronavirus. In: Pandemic detection and analysis through smart computing technologies. pp 195–224

World Health Organization (2022) Weekly epidemiological update on COVID-19-7 September 2022, Edition 108

Yang W, Cao Q, Qin L, Wang X, Cheng Z, Pan A, Dai J, Sun Q, Zhao F, Qu J, Yan F (2020a) Clinical characteristics and imaging manifestations of the 2019 novel coronavirus disease (COVID-19): A multi-center study in Wenzhou city, Zhejiang, China. J Infect 80(4):388–393

Yang L, Liu W, Yu X, Wu M, Reichert JM, Ho M (2020b) COVID-19 antibody therapeutics tracker: a global online database of antibody therapeutics for the prevention and treatment of COVID-19. Antibody Therapeutics 3(3):205–212

Performance Analysis of Matrix Multiplication Based on Serverless Platform

Junzhou Chen, Jiren Lu, and Hanpei Ma

Abstract Over the past few years, serverless computing has become increasingly important as a new field that dramatically reduces costs, reduces latency, improves scalability, and eliminates server-side management. Big data is also increasingly used in various areas, such as finance, transportation, and telecommunications. Large amounts of data information must be stored in high-dimensional matrices for large-scale scientific computations. This paper combined serverless computing with high-dimensional matrices by using the MapReduce model to implement a serverless architecture on Aliyun to multiply high-dimensional matrices. This paper used observation, comparison, and control variable methods to conduct experiments and concluded by analyzing a large amount of experimental data. This paper found that the change in the number of workers will affect the total working time, the ratio among total working time, calculation time and communication time. The matrix size change will affect the computation time ratio to total working time and communication time to total working time. Moreover, this paper infers an optimal number of workers for each size matrix, which minimizes the total execution time.

J. Chen, J. Lu, and H. Ma these authors contributed equally.

J. Chen
Westa College, Southwest University, Chongqing 400715, China

J. Lu (✉)
School of Statistics and Information, Shanghai University of International Business and Economics, Shanghai 201620, China
e-mail: 20057010@suibe.edu.cn

H. Ma
School of Information and Computer, Taiyuan University of Technology, Taiyuan 030000, China

S. Yadav et al. (eds.), *Proceedings of 2nd International Conference on Artificial Intelligence, Robotics, and Communication*, Lecture Notes in Electrical Engineering 1063, https://doi.org/10.1007/978-981-99-4554-2_11

1 Introduction

From IaaS (infrastructure as a service) to PaaS (platform as a service) and then to SaaS (software as a service), cloud computing continues to evolve uninterruptedly, and the trend of FaaS (Function as a Service) is becoming more and more apparent (Hassan et al. 2021). To date, serverless computing has been the most favored execution model for cloud computing, with the most significant benefit of providing a distributed, flexible, and scalable computing execution environment that only pays for the actual use of resources. With serverless, customers can be freed from regular operation and maintenance matters, and it is more convenient for them to focus on specific tasks (Jonas et al. 2019). The application of serverless computing as a cloud-native model has become diversified, and its related products have become a competing item for various cloud providers. Serverless computing was introduced by the Amazon cloud provider in 2014 at the re: Invent conference "Getting Started with AWS Lambda" with the first serverless computing platform AWS Lambda. After Amazon launched AWS Lambda, other cloud providers have also launched serverless computing platforms, including Google Cloud Functions, Microsoft Azure Functions, IBM OpenWhisk, and Alibaba Aliyun (Baldini et al. 2017).

Serverless computing is a cloud service, moreover, a new architectural paradigm for software computing, also known as Functions as a Service (FaaS), whose name does not imply the absence of servers, but—rather—the absence of consideration (or concern) for servers, meaning that customers do not need to worry about server management. A serverless platform that allows small pieces of code to run for a limited period, whose execution is triggered by events or HTTP (or other triggers), allows customers to run and manage application code against triggered functions but does not allow them to maintain a persistent state, as functions can be restarted at any time. Customers deploy small code units to FaaS, and the platform runs the code on customer demand, scaling automatically (Castro et al. 2019). While allowing customers to run event-driven functions in the cloud, they do not need to manage their resource allocation or configure the runtime environment (Jangda et al. 2019). This is because the platform's functional services are provided by the FaaS provider, such as significant maintenance of servers, security updates, availability, reliability monitoring, and troubleshooting (Castro et al. 2019). The platform fully manages the operating system and how resources are allocated. The provider allocates sufficient resources in real-time, and customers do not need to worry about server management (Werner et al. 2018). Customers run code on serverless platforms and only pay for the code that they run. There is no cost involved in deploying serverless cloud computing infrastructure and building applications (Rahman and Hasan 2019).

MapReduce is a programming model for processing and generating parallel operations on large-scale datasets, which can be divided into "Map" and "Reduce" (Condie et al. 2010). The user specifies a Map function to map a set of key-value pairs into a new set of key-value pairs, and a concurrent Reduce function to ensure that each of the mapped key-value pairs shares the same set of keys. It's not as popular as it was before, but it's still very important and even now it provides great value (Dean and Ghemawat 2010).

In the considerable data age, data comes from e-mail, video, audio, images, and social network interactions. Data information needs to be stored in high-dimensional matrices for large-scale scientific calculations. It becomes difficult to capture, form, store, manage, share, analyze, and visualize through specific database software tools (Sagiroglu and Sinanc 2013). More and more attention has been paid to the statistics and reasoning of high-dimensional data, which has attracted significant interest from information science researchers, government and enterprise policymakers, and decision-makers (Chen and Zhang 2014). Matrix multiplication is a common bottleneck in scientific computing, machine learning, graphics processing and other fields. In many applications, this matrix is huge, and its size can easily be extended to millions. The cost of moving data between nodes is now several orders of magnitude higher than the cost of arithmetic operations. Over time, this gap grows exponentially, leading to the popularity of parallel computing algorithms that avoid communication (Gupta et al. 2018).

This paper, based on the hotness and prospect of serverless computing and matrix multiplication operations, referring to the MapReduce framework, constructs a model which can realize high-dimensional matrix multiplication based on Aliyun serverless function calculation and object storage OSS (Object Storage Service) platform. This paper specifies two matrices of a certain size to perform the multiplication and the number of workers to call. The native function then generates these two matrices with random elements of defined size, divided equally by the number of worker threads, and uploads them to OSS for storage. The control function will then call the appropriate number of workers to complete the calculation. Finally, the control function receives the completed data from the workers, integrates the data, and returns the complete matrix multiplication result. This paper first keeps the matrix size unchanged when conducting experiments and changes the number of worker threads to test the experimental data under different computing nodes. Then the number of workers is kept constant, and the matrix size is changed to test the experimental data under different matrix sizes. This paper found that when the matrix size was constant, and the number of workers was increased, the total working time of the program first decreased and increased later. As the number of workers increased, the proportion of computation time to communication time kept decreasing, the ratio of computation time to total working time kept decreasing, and the proportion of communication time to total working time kept increasing. When the number of workers was constant, and the matrix size increased, the optimal number of workers per matrix kept increasing. As the matrix size increased, the ratio of computation time to total working time rose, and the proportion of communication time to total working time kept decreasing.

2 Framework Design

This paper referred to the MapReduce Framework and built a model that can accomplish high-dimensional matrix multiplication based on the Aliyun serverless function computation and object storage OSS platform. The flow demonstration diagram is shown in Fig. 1.

First, this paper specifies two matrices of definite size to do the multiplication operation and specify the number of invoked workers. Then the native function generates these two matrices of definite size with random elements inside the matrix. The matrices are split equally by the number of workers and uploaded to OSS for storage.

Second, the control function will invoke the corresponding number of workers to complete the calculation. The worker function receives the instructions sent by the Control function and goes to the OSS to read the data that needs to be used, which will be calculated and returned to the Control function for the calculation result.

Third, the control function receives the completed data from the workers, integrates the data, and returns a complete matrix multiplication result, which is the result of the multiplication of two matrices randomly generated by the native program.

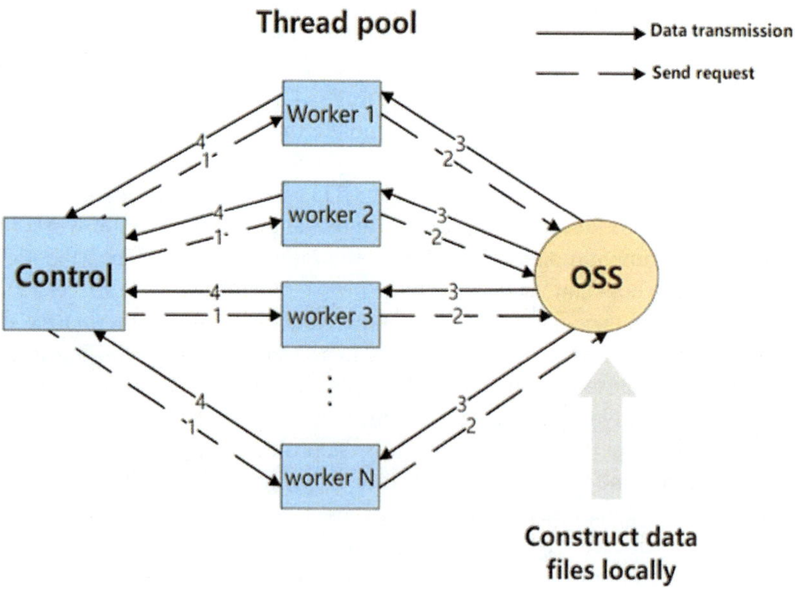

Fig. 1 Model flow demonstration diagram

Table 1 randomMatrix

Algorithm 1. randomMatrix

Input: m,n // size of matrix
Output: Matrix

```
Matrix ← a new list [ ]
for i ← 0 to m do:
    L← a new list [ ]
    for j ← 0 to n do:
        L.append (a random number between 0 and 255)
    end for
    Matrix.append (L)
end for
return Matrix
```

3 Function Description

3.1 *Native Function*

The Native function consists of randomMatrix (Table 1) and creatDataFile (Table 2). The randomMatrix function will create two matrices, A and B, with certain sizes but random elements, which simulate the 8-bit photos matrix by covering 0–255 elements. Then the creatDataFile function will split the two matrices equally according to the number of workers, and the split process is as follows (Fig. 2).

For example, this paper assumes that the quantity of workers is 16, and each matrix is divided into four parts. Matrix A is split horizontally, matrix B is split vertically, and the split matrix is marked with "m" and "n" (as shown in Fig. 2) to facilitate reading data in subsequent calculations. The split matrix is converted into a JavaScript Object Notation (JSON) format data file to transfer data. The naming of the file contains information about which matrix and block (for example, the first block of matrix A is named "matrix_A_0"), then upload these files to OSS for storage. The function randomMatrix (Table 1) and creatDataFile (Table 2) codes are as follows.

3.2 *Control Function*

The Control function calls the Worker functions, allocates work, and accepts and integrates the calculation results. The Control function consists of getResultFromAliyun (Table 3) and mainControlFunction (Table 4).

The mainControlFunction will first generate the dictionary of the matrix splitting information, including the coordinates "m" and "n" for communication to the Worker

Table 2 creatDataFile

Algorithm 2 creatDataFile

Input: i, j, k, split_num
A ← matrix(i, j)
B ← matrix(j, k)
bucket = oss2.Bucket (auth, endpoint, bucket_name)
for a ← 0 to split_num **do**:
    mat_A ←Start the A at a * (i \ split_num), and slice to position (a + 1) * (i \ split_num)
    store mat_A in json format named A_a.json
    use bucket to put file A_a.json as matrix_A_a in OSS
end for
for b ← 0 to split_num **do:**
    mat_B ←a new list [ ]
    for line ← each row in B **do:**
        mat_B.append(Start the line at b * (k \ split_num), and slice to position (b + 1) * (k \ split_num))
    end for
    store mat_B in json format named B_b.json
    use bucket to put file B_b.jsom as matrix_B_b in OSS
end for

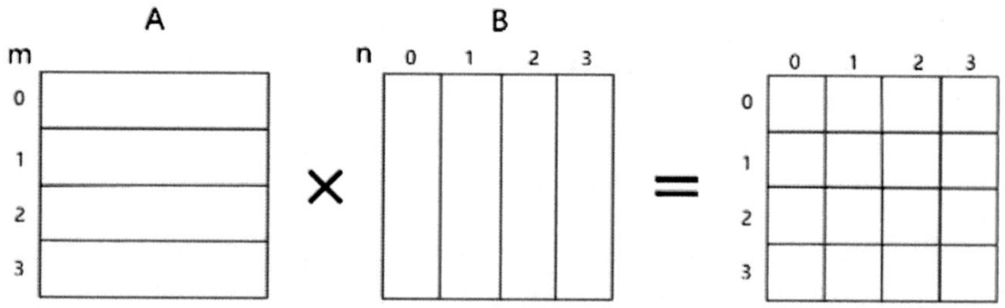

Fig. 2 Split process diagram

Table 3 getResultFromAliyun

Algorithm 3 getResultFromAliyun

Input: data
Output: res.json()
try:
    session ← data['session']
    params ← {'m': data['m'], 'n': data['n']}
    res ← session.post(url, data ← params)
    return res.json()
except Exception as e:
    print (e)

Table 4 mainControlFunction n

Algorithm 4 mainControlFuction

```
Input: i, j, k, split_num
splitResult ←a new list [ ]
session ← requests.session()
for m ← 0 to split_num do:
    for n ← 0 to split_num do:
        splitResult.append({'m': m, 'n': n,'session':session})
    end for
end for
executor ← ThreadPoolExecutor(max_workers ← split_num²)
result ←a new dict { }
start ← current time
cal_time ← 0
for data in executor, map (get_result_from_aliyun, splitResult) do:
    if data.get('m') ← -999 then:
        print (data, get ('result'))
        break
    end if
    result[(int(data.get('m')), int(data.get('n')))] ← data.get('result')
    cal_time ← data.get('time') + cal_time
end for
print(current time - start)
res ← a new list [ ]
for m in range(split_num) do:
    for line ← 0 to length of result[(0, 0)] do:
        L ←a new list [ ]
        for n ← 0 to split_num do:
            L.extend(result[(m, n)][line])
        end for
        res.append(L)
    end for
end for
output res as the result of computation
output cal_time / split_num² as time consumption of computation
```

function about the matrix position information. The Control function places these dictionaries in a List, then uses multi-threading to traverse the List and call the getResultFromAliyun function. When added to the thread pool, the getResultFromAliyun function will call the Worker function and receive the calculation results. The calculation results will be returned to the primary function and put into a dictionary. In the dictionary, the key is the coordinates (m, n) of the matrix block, and the value is the 2-dimensional List of the calculation results. In this process, the function will record and output the time from calling the getResultFromAliyun function to receiving all calculation results. At last, the program will read the List of corresponding coordinates from the dictionary and merge it into the output result, then the function ends. The function getResultFromAliyun (Table 3) and mainControlFunction (Table 4) codes are as follows.

3.3 Worker Function

The Worker function consisting of matrixMultiplication (Table 5) and matrixDownload (Table 6) computes the node function and is executed concurrently. The maximum number of concurrent instances in this paper set is 100. The matrixMultiplication will receive the coordinates (m, n) of matrix blocks A and B and download the corresponding matrix data from OSS using the matrixDownload function. If the download fails, the matrixDownload function will sleep for one second and download again to prevent excessive OSS requests. If the download fails more than five times, it will cancel the download work and return an empty list. In the function "randomId = math.ceil(time.time())" create a random ID to prevent reading and writing failures caused by multi-threaded files with the same names.

NumPy (Numerical Python) is used for matrix calculation, and the calculation result is converted into a two-dimensional list by "tolist()" method. The time used for Numerical Python calculation and "tolist()" conversion is recorded here. This time will also be returned to the Control function, which requires all threads to calculate

Table 5 matrixMultiplication

Algorithm 5 matrixMultiplication

Input: request
Output: json file

```
try:
    m ← request.form['m']
    n ← request.form['n']
    randomld ←Round number of current time upward to its nearest integer
    mat_A_Json ← download(matrix_A_in, a_m_randomld.json)
    mat_B_json ← download(matrix_B_n, b_n_randomld.json)
    timel ← current time
    mat ← np.matmul(mat_A_Json,mat_B_Json).tolist()
    time2 ← current time
    return jsonify({'m': m,'n': n,'time': time2-time1,'Result': mat})
except Exception as e:
    return jsonify({'m':-999,'n': n,'time': 0,'Result':e})
```

Table 6 matrixDownload

Algorithm 6 matrixDownload

```
Input: remoteName, localName, retry = 0
Output: mat_json
mat_json ← a new list [ ]
try:
    bucket.get_object_to_file(remoteName, localName)
    mat_json ← read file named localName
except Exception as e:
    print (e)
if length of mat_json <= 0 then:
    if retry > 5 then:
        return an empty list
    sleep for 1 second
    mat_json ← download(remoteName, localName, retry + 1)
end if
return mat_json
```

the average time. The Worker function will return a dictionary to the Control function, including the coordinates of blocks "m" and "n," the calculation time, and the calculation results. The function matrixMultiplication (Table 5) and matrixDownload (Table 6) codes are as follows.

4 Experimental Data

This paper completed the high-dimensional matrix multiplication operation with Alibaba Cloud's Serverless function computation and object storage OSS. The worker is set to a maximum of 100 instances concurrently.

First, the matrix size remains unchanged, and the number of workers is changed to test the experimental data under the different number of computation nodes. Because the worker number is the square of a spilled number, this paper only did experiments with worker numbers 4, 9, 16, 25, 64, and 100. To get as much data as possible for the worker number group, this paper selected the matrix size of 1200 × 1200 to experiment with. The experimental results are given in Table 7.

Then the worker number remains unchanged, and the matrix size is changed to test the experimental data under different matrix sizes. The number of workers is set to 16, and the matrix sizes are set to 400 × 400, 800 × 800, 1200 × 1200, and 1600 × 1600. This paper also tested the 8000 × 8000 and 12,000 × 12,000 size matrices with several workers 16, and the 8000 × 8000 size matrices each took more than 30 min to compute. The 12,000 × 12,000 size matrix consumed more memory per worker than the Ali cloud function's maximum memory usage of 32 GB. The experiment

Table 7 Change the number of workers (size of matrix: 1200 × 1200)

Number of workers	4	9	16	25	64	100	
Memory of each matrix after splitting (MB)	3.139	2.093	1.57	1.256	1.046	0.6425	
Control function	Execution time (s)	9.846	4.28	4.426	4.619	13.35	19.98
	Memory consumption (MB)	143.03	143.53	131.01	151.52	161.54	210.95
	Time consumption of multi-threaded computation (s)	9.7739	3.4087	4.3338	4.8358	13.25	19.0124
Worker function	Execution time (average) (s)	8.684	2.764	4.24	4.121	12.425	18.49
	Computation time (average) (s)	6.3332	0.6646	0.5275	0.2307	0.4804	0.1703
	Memory consumption (MB)	425.77	308.96	258.17	251.8	303.25	342.04
Data analysis	Communication time (s)	3.441	2.7441	3.8063	4.6050	12.7740	18.8420
	Computation time/communication time	0.6479	0.1949	0.1217	0.0477	0.0362	0.0089
	Computation time/total time	0.6432	0.1552	0.1191	0.0499	0.0359	0.0085
	Communication time/total time	0.3494	0.6411	0.8599	0.9969	0.9568	0.9430

was unsuccessful and more workers should be called to do the calculation. The experimental results are given in Table 8. From the experimental data, this paper can calculate the following data: (1) Communication time (time consumption of multi-threaded calculation subtract time consumption of calculation). (2) Computation time/Communication time (the function execution time and communication time ratio). (3) Computation time/Total time. (4) Communication time/Total time.

By observing and comparing the experimental data: (1) Using data in Table 7, the matrix size was constant (1200 × 1200), and the number of workers increased (from 4 to 100). The total working time of the program (execution time) decreased at first and increased later. (2) By using data in Table 7, as the number of workers increased (from 4 to 100), the proportion of computation time to communication time kept decreasing, the ratio of computation time to total working time kept decreasing, and the proportion of communication time to total working time kept increasing. (3) By using data in Table 8, as the matrix size increased (from 400 × 400 to 4000 × 4000), the ratio of computation time to total working time (computation time/total time) rose, and the proportion of communication time to total working time (communication time/total time) kept decreasing.

Table 8 Change the size of matrix (number of workers: 16)

Size of matrix	400 × 400	800 × 800	1200 × 1200	4000 × 4000	
Memory of each matrix after splitting (MB)	0.1785	0.7144	1.57	17.435	
Control function	Execution time(s)	1.507	4.175	4.426	329.433
	Memory consumption (MB)	71.57	100.84	131.01	1060
	Time consumption of multi-threaded computation(s)	1.4853	4.1342	4.3338	328.7798
Worker function	Execution time (average) (s)	1.117	3.22	4.24	318.997
	Computation time (average) (s)	0.0114	0.2566	0.5275	257.7049
	Memory consumption (MB)	83.91	135.9	258.17	2430
Data analysis	Communication time (s)	1.4739	3.8775	3.8063	71.0748
	Computation time/ communication time	0.0076	0.062	0.1217	0.7838
	Computation time/total time	0.0075	0.0614	0.1191	0.7822
	Communication time/ total time	0.978	0.9287	0.8599	0.2157

This paper extrapolated from the data and the first conclusion above that there is an optimal number of workers for each size matrix to minimize the total execution time, so this paper did the following experiment. This paper changes both matrix size and worker number by testing the optimal number of workers for each matrix size. The experimental data are as Table 9. The blank space in the table is not possible (this model cannot be allocated evenly to workers to compute) or unnecessary (monotonous in range) for experiments.

The experimental data in Table 9 and Fig. 3 can be drawn to observe the data more directly. Each line represents a matrix of the same size, varying the total operation time as the number of workers increases. Matrices of different sizes are represented by lines of different colors. When the matrix size is 400 × 400 or 800 × 800, the optimal number of workers is less than or equal to four. When the matrix size is 2400 × 2400, the optimal number of workers is greater than or equal to 100.

From Fig. 3 can draw the conclusion that the number of workers was constant and the size of matrix increased, the optimal number of workers per matrix size kept increasing.

Table 9 Time consumption for different matrix sizes and number of workers

Size of matrix	Number of workers						
	4	9	16	25	36	64	100
400 × 400	0.464	–	1.6	1.827	2.79	–	–
800 × 800	1.754	–	2.38	3.319	–	4.493	5.492
1200 × 1200	9.846	4.28	4.42	4.619	–	13.35	19.98
1600 × 1600	16.797	–	9.677	12.284	–	14.292	–
2000 × 2000	–	–	39.661	25.979	–	33.659	–
2400 × 2400	–	–	–	–	65.677	49.777	38.255

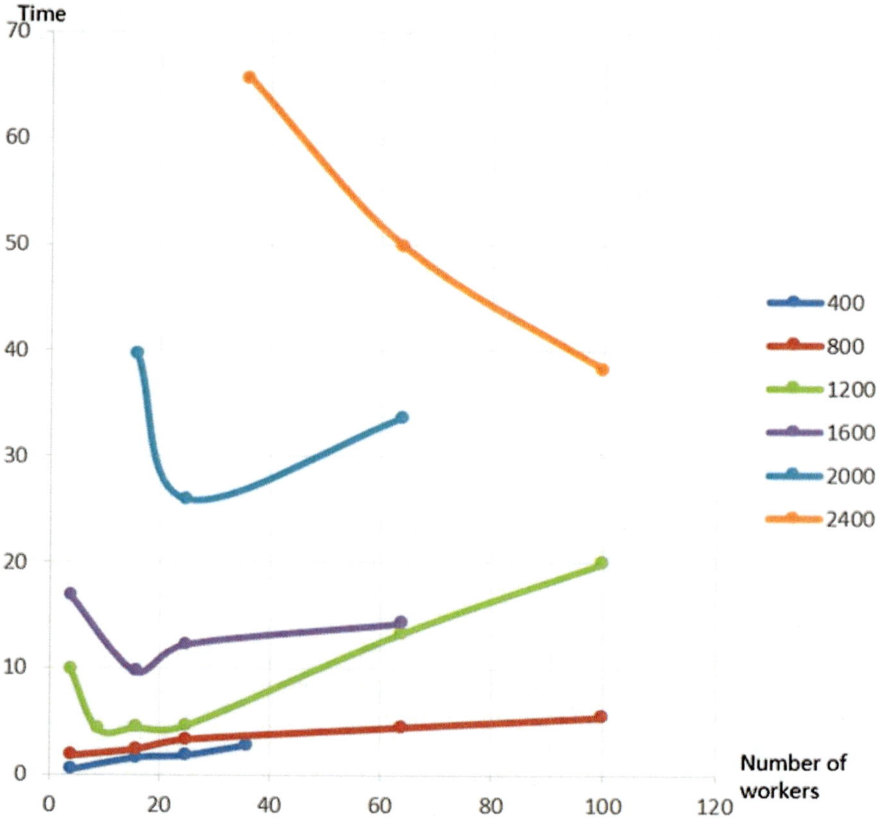

Fig. 3 Time consumption for different matrix sizes and number of workers

5 Discussion

This paper's experiments have the limitation that the scheme in this paper can only assign the matrix to a fixed number of computational nodes (4, 9, 16, 25, 36, 64 ...), and this paper hope that future research will design an algorithm so that the matrix can be evenly partitioned. Any number of computational nodes can be called for computation. This paper found that although the matrix size assigned to each computational node is the same, there is a difference in their running time, and this paper hopes that future research will identify the reason for this and improve it, or can dynamically allocate the computational volume according to the computational speed of the unused nodes. This paper would like to improve the model by reducing the total time consumption using a compute-while-communicating approach or by reducing the communication time and polymerizing time concerning other distributed computing approaches. All of this paper's experimental conclusions are based on observations of the experimental results, and this paper hopes that future studies will produce more precise results based on mathematical methods.

6 Conclusion

This paper explored serverless infrastructure for big data processing using matrix multiplication as an example—this paper designed and implemented a computing framework using Aliyun. Specify the matrix size, the number of workers and the native program randomly generated matrix. Average splitting is completed and uploaded to the OSS storage. The control function invokes the Worker function to read the data in the OSS and calculate. The results will be returned to the Control function, the Control function to integrate the data and output the complete calculation results. Through the model, this paper studies the relationship among matrix size, number of workers, and execution time and gets several conclusions. The constant size of the matrix and the increase in the number of workers will lead to a decrease in the total run time before an increase, a decrease in the computation time of a single worker, and an increase in the time spent on data transfer, it is deduced that the ratio of computation time to communication time is decreasing, while the ratio of communication time to total working time is increasing. As the number of workers increases, the total elapsed time decreases first and then increases. There is an optimal number of workers for each matrix size and the optimal number of workers increases as the matrix size increases. The constant number of workers and the increase in the size of the matrix lead to an increase in the computation time of a single Worker. While the time spent on data transfer is also increasing, the proportion of computation time to total run time increases, and communication time to total running time decreases.

References

Baldini I, Castro P, Chang K, et al (2017) Serverless computing: current trends and open problems. In: Research advances in cloud computing. Springer, Singapore, pp 1–20

Castro P, Ishakian V, Muthusamy V et al (2019) The rise of serverless computing. Commun ACM 62(12):44–54

Chen CP, Zhang CY (2014) Data-intensive applications, challenges, techniques and technologies: a survey on Big Data. Inf Sci 275:314–347

Condie T, Conway N, Alvaro P, et al (2010, April) MapReduce online. In: Nsdi, vol 10, No 4. p 20

Dean J, Ghemawat S (2010) MapReduce: a flexible data processing tool. Commun ACM 53(1):72–77

Gupta V, Wang S, Courtade T, et al (2018, December) Oversketch: Approximate matrix multiplication for the cloud. In: 2018 IEEE international conference on big data (Big Data). IEEE, pp 298–304

Hassan HB, Barakat SA, Sarhan QI (2021) Survey on serverless computing. J Cloud Comput 10(1):1–29

Jangda A, Pinckney D, Brun, et al (2019) Formal foundations of serverless computing. In: Proceedings of the ACM on programming languages, vol 3 (OOPSLA). pp 1–26

Jonas E, Schleier-Smith J, Sreekanti V, et al (2019) Cloud programming simplified: a Berkeley view on serverless computing. arXiv preprint arXiv:1902.03383

Rahman MM, Hasan MH (2019, October) Serverless architecture for big data analytics. In: 2019 global conference for advancement in technology (GCAT). IEEE, pp 1–5

Sagiroglu S, Sinanc D (2013) Big data: a review. In: 2013 international conference on collaboration technologies and systems (CTS). IEEE, pp 42–47

Werner S, Kuhlenkamp J, Klems M, et al (2018) Serverless big data processing using matrix multiplication as example. In: 2018 IEEE international conference on Big Data (Big Data). IEEE, pp 358–365

Pilot Design for Compressed Sensing Based OFDM Channel Estimation

Shi Liu and Ping Li

Abstract The channel estimation technique of classical pilot-based leads to better signal reconstruction performance. However, most communication systems ignore the adopted multipath assumption, and the channels exhibit sparsity. Channel sparsity is exploited by compressed sensing. The results show that in order to obtain a good sensing matrix, a proper design of the measurement matrix with respect to the pilot pattern is required. A proposed Random Search Algorithm is proposed to search for the optimal pilot location and use an appropriate nonlinear reconstruction algorithm. By comparing the conventional channel estimation with the evenly pilot pattern, the results show that there are different degrees of improvement in the channel estimation performance.

1 Introduction

Using techniques from transform domain coding, compressed sensing has emerged as a new framework for signal acquisition and sensor design, which allows signals with sparse and compressive representation properties to be sampled and computed at a much lower cost. The Nyquist-Shannon sampling theorem states that to perfectly acquire an arbitrary band-limited signal, then a specific minimum number of samples is required (Baraniuk 2007), but when the signal is known to be sparse, then the number of samples required can be significantly reduced, thus reducing data storage. Thus, better results can be obtained than traditional methods when sensing sparse signals, such as reducing the number of samples while maintaining the quality of signal recovery. The basic idea behind compressed sensing is the preference to find ways to sense data directly from a compressed format, i.e., using a lower sample rate, rather than first sampling at a high sample rate and then compressing the sampled data. Compressed sensing methods have the advantages of less lead time, high accuracy,

S. Liu (✉) · P. Li
Department of Information Science and Engineering, Dalian Polytechnic University, Dalian, China
e-mail: ls306070221@163.com

S. Yadav et al. (eds.), *Proceedings of 2nd International Conference on Artificial Intelligence, Robotics, and Communication*, Lecture Notes in Electrical Engineering 1063, https://doi.org/10.1007/978-981-99-4554-2_12

and high spectrum utilization, and have been widely used in sparse channel estimation (Candes 2006; Candes and Romberg 2006). Unlike the conventional channel estimation schemes, the evenly-pilot mode is not suitable for channel estimation. For signal reconstruction, a better channel estimation can be obtained by using random-pilot mode.

2 System Modeling

Broadband wireless communication channels usually have multipath effects, in which the transmitted signal waveform will suffer from frequency selective fading due to simultaneous amplitude and phase distortions. In such a channel, it is assumed that the impulse response of the channel does not change with time within an OFDM symbol. At this point, a sparse multipath channel of length $\boldsymbol{Lh} = \left[\boldsymbol{h}_0, \boldsymbol{h}_1, \ldots, \boldsymbol{h}_{L-1}\right]^{\mathrm{T}}$ has a time domain impulse response of:

$$\boldsymbol{h}_i = \sum_{j=1}^{L} \boldsymbol{a}_j \delta\left(\boldsymbol{i} - \boldsymbol{d}_j\right)(0 \leq \boldsymbol{i} \leq \boldsymbol{L} - 1) \tag{1}$$

where $\boldsymbol{a}_j$ is the complex gain of the $\boldsymbol{j}$th path and $\boldsymbol{d}_j$ is the time delay of the $\boldsymbol{j}$th path. If only $\boldsymbol{K}$ of these channel taps have non-zero values and $\boldsymbol{K} \ll \boldsymbol{L}$ then this channel is said to be sparse, and its sparsity is $\boldsymbol{K}$.

Assuming that there are $\boldsymbol{N}$ subcarriers in an OFDM system and that the length of the cyclic prefix added before its symbol is larger than the maximum delay extension of the channel, the OFDM signal after transmission through the wireless channel, after removing the cyclic prefix and performing the $\boldsymbol{N}$-point DFT transform at the receiver side can be expressed as:

$$\boldsymbol{Y} = \boldsymbol{X}\boldsymbol{H} + \boldsymbol{N} \tag{2}$$

where, $\boldsymbol{Y} = \left[\boldsymbol{Y}_0, \boldsymbol{Y}_1, \ldots, \boldsymbol{Y}_{L-1}\right]^{\mathrm{T}}$, $\boldsymbol{X} = \mathbf{diag}[\boldsymbol{X}(0), \boldsymbol{X}(1), \ldots, \boldsymbol{X}(\boldsymbol{N}-1)]$; $\boldsymbol{H} = \boldsymbol{W}_{\boldsymbol{N}\times\boldsymbol{L}}\boldsymbol{h}$ is the frequency domain value of the channel; $\boldsymbol{W}_{\boldsymbol{N}\times\boldsymbol{L}}$ is the partial discrete Fourier transform matrix:

$$\boldsymbol{W} = \frac{1}{\sqrt{\boldsymbol{N}}}\begin{pmatrix} \boldsymbol{w}^{00} & \cdots & \boldsymbol{w}^{0(\boldsymbol{L}-1)} \\ \vdots & \ddots & \vdots \\ \boldsymbol{w}^{(\boldsymbol{N}-1)0} & \cdots & \boldsymbol{w}^{(\boldsymbol{N}-1)(\boldsymbol{L}-1)} \end{pmatrix} \tag{3}$$

where $\boldsymbol{w}_{\boldsymbol{N}}^{\boldsymbol{nl}} = \frac{1}{\sqrt{\boldsymbol{N}}}\boldsymbol{e}^{-j\frac{2\pi\boldsymbol{nl}}{\boldsymbol{N}}}$, $\boldsymbol{n} = 0, 2, \ldots, \boldsymbol{N} - 1, \boldsymbol{l} = 0, 2, \ldots, \boldsymbol{L} - 1$. $\boldsymbol{N}$ is the $\boldsymbol{N}$-dimensional complex Gaussian white noise with variance of $\boldsymbol{\sigma}^2$.

If $\boldsymbol{P}$ subcarriers among $\boldsymbol{N}$ subcarriers of OFDM are selected to transmit the pilot symbols, the received pilot signal is:

$$\boldsymbol{Y_P} = \boldsymbol{X_P W_P h} + \boldsymbol{N_P} = \boldsymbol{Ah} + \boldsymbol{N_P} \tag{4}$$

where the $\boldsymbol{P} \times 1$-dimensional vector $\boldsymbol{Y_P}$ is the received pilot signal; the $\boldsymbol{P} \times \boldsymbol{P}$-dimensional matrix $\boldsymbol{X_P}$ is a diagonal matrix whose elements on the diagonal are the $\boldsymbol{P}$ pilot symbols sent; the $\boldsymbol{P} \times \boldsymbol{L}$-dimensional matrix $\boldsymbol{W_P}$ is a discrete Fourier transform array with only the pilot rows selected; and the $\boldsymbol{P} \times 1$ vector $\boldsymbol{N_P}$ is the channel noise corresponding to the pilot rows. $\boldsymbol{Y_P}$, $\boldsymbol{X_P}$, $\boldsymbol{W_P}$ are known at the receiver side.

In Eq. (4), $\boldsymbol{W_P}$ is equivalent to the basis matrix and $\boldsymbol{X_P}$ is equivalent to the observation matrix, so it is possible to obtain the time-domain response value $\boldsymbol{h}$ of the channel using the compressed sensory reconstruction algorithm, and then substitute the obtained $\boldsymbol{h}$ into Eq. $\boldsymbol{H} = \boldsymbol{Wh}$ to obtain the frequency-domain response value $\boldsymbol{H}$ of the channel.

3 Pilot Design

For a given signal $x \in \mathbb{R}^n$, consider a measurement system that obtains m linear measurements, in a mathematical sense this description can be expressed as:

$$y = Ax \tag{5}$$

where A is a matrix of order $m \times n$ and $y \in \mathbb{R}^m$. The matrix A represents a dimensionality reduction matrix, i.e., a mapping from $\mathbb{R}^n$ to $\mathbb{R}^m$. Since $n > m$ for most compressed sensing scenarios, the system is defined as an underdetermined system, i.e., there are more unknowns than observations. So, it is usually not possible to obtain an exact reconstruction of the original input x using the conventional inverse transform of A. Therefore, if there is a priori information about the sparsity of the signal and a condition is imposed on A, then the reconstruction can be performed by solving the l_1 parametric minimization problem as follows:

$$\min_{\mathbf{x}} \|\mathrm{x}\|_1 \text{ subject to } Ax = y \tag{6}$$

When observations are polluted by noise or corrupted by errors when quantizing, some stronger conditions must be considered. Matrix A satisfies the RIP (Hwang et al. 2008; Wei et al. 2014; Candes et al. 2006) condition of order k, if there is a constant $\delta_k \in (0, 1)$, so that:

$$(1 - \delta_k)\|x\|_2^2 \leq \|Ax\|_2^2 \leq (1 + \delta_k)\|x\|_2^2 \tag{7}$$

For any k-sparse vector x with uniform constant δ_k if it satisfies $\delta_{2k} < \sqrt{2}-1$. Meanwhile, if the matrix A satisfies the RIP condition, then it is sufficiently guaranteed to reconstruct the sparse signal from the observations with noise, so the RIP condition is necessary if all sparse signals x are reconstructed from the observations Ax. RIP provides a guarantee to restore k sparse signal. To determine a matrix A meets the condition, search all $\binom{n}{k}$ submatrix. In most cases, the more computable conditions of matrix A are used to give more specific reconstruction guarantees. The coherence of a matrix is one of these properties (Donoho and Elad 2003; Tropp and Gilbert 2007).

Definition 1 The coherence $\mu(A)$ of matrix A is defined as the inner product of the maximum absolute value of any two columns a_i, a_j of A (Eladar 2009).

$$\mu(A) = \max_{1 \le i < j \le n} \frac{|\langle a_i, a_j \rangle|}{\|a_i\|2\|a_j\|_2} \tag{8}$$

The coherence of the pilot position can be expressed as:

$$\begin{aligned}\mu(A) &= \max_{0 \le i < j \le L-1} \left| \sum_{i=0}^{\mathcal{P}} X(K_i) e^{-j2\pi k_i m})/N X(K_i) e^{-j2\pi k_i n})/N \right| \Bigg/ \left(\sum_{i=0}^{\mathcal{P}} |X(K_i)|^2 \right) \\ &= \max_{0 \le i < j \le L-1} \left| \sum_{i=0}^{\mathcal{P}} |X(K_i)|^2 e^{j2\pi k_i (n-m)/N} \right| \Bigg/ \left(\sum_{i=0}^{\mathcal{P}} |X(K_i)|^2 \right)\end{aligned} \tag{9}$$

Therefore, the coherence of the measurement matrix A for the representation channel estimation objective function is:

$$Q =: \min_{\mathcal{P}} \mu(A) \tag{10}$$

From Eq. (9), the optimal pilot mode can be obtained by:

$$\mathcal{P}_{\text{opt}} = \operatorname{argmin} \mu(A) \tag{11}$$

4 Random Search Algorithm

For a linear system of equations $y = Ax$, compression perception tends to find the unique sparse l_1 parametric solution. The problem is expressed as:

$$\arg \min_{\hat{\hat{x}}} \|\hat{x}\|_p \text{ s.t. } y = A\hat{x} \tag{12}$$

where x is the k-sparse channel impulse response, A is the measurement matrix, $p = 1$, $\|x\|_{l_1} = \sum_{k=1}^{N} |x_k|$ is the l_1 parametric relaxation of the l_0 parametric number, and y belongs to the observed signal. A valid measurement matrix is guaranteed if the derivative in position $\mathcal{P}_i$ is at the position where the maximum absolute inner product between columns of the measurement matrix is smaller. The proposed random search algorithm using the criterion in Eq. (11) and search for the best lead frequency position using this criterion is used. The specific steps are as follows:

Step 1: Generate T subsets of the random derivative subcarrier M such that: $\mathcal{P}_t \in N, t = 1, 2, \ldots, T$ and $\mathcal{P} = \{K_1, K_2, \ldots, K_{\mathcal{P}}\}$.

Step 2: Determine the consistency between the columns of the measurement matrix, as shown below:

$$\mu_t(A) = 1. \max_{0 \le i < j \le L-1} \frac{1}{\mathcal{P}} \left| \sum_{i=0}^{\mathcal{P}} e^{j2\pi \mathcal{P}_i (n-m)} \right| \tag{13}$$

Step 3: The minimum coherence of step 2 is determined according to Eq. (12). Thus, $\mu_r(A) = \min\{\mu_t(A)\}$, where $r \in T, t = \{1, 2, \ldots, T\}$ and $r = \text{opt}(\mathcal{P}_r = \mathcal{P}_{\text{opt}})$.

5 Simulation Results and Analysis

5.1 Mean Square Error Versus BER in Channel Estimation

In order to verify that random-pilot mode usually has better performance than the continuous-pilot mode and the evenly-pilot mode in compressed sensing channel estimation, the following two pilots modes are used in this paper: CS-based continuous-pilot mode; CS-based evenly-pilot mode, where the system parameter model parameters are set as shown in Table 1.

Figure 1 Experiment shows that the evenly and random pilot performance of CS-based channel estimation is better than that of least square channel estimation. Meanwhile, the stochastic pilot mode is better than the continuous and evenly-pilot mode, and the continuous-pilot mode has the worst performance. The evenly-pilot mode selected in the traditional channel estimation algorithm is no longer advantageous

Table 1 System model simulation parameters for verifying performance in different conduction modes

Parameter name	Numerical value
Pilot space	21
Number of subcarriers	512
Number of pilots	24
Channel length	50

in CS-based channel estimation scheme. Figure 2 shows the BER plots for OFDM systems in three cases: CS-based channel estimation using different pilots, Least Square channel estimation using evenly pilots, and a comparison using complete channel state information (Xia et al. 2004). It can be seen that the Least Square channel estimation is weaker for system performance improvement than the CS-based channel estimation under continuous-pilot mode. The Bit Error Ratio of the proposed random-pilot mode is close to the BER with full CSI, which proves the efficiency of random-pilot mode.

Fig. 1 The relationship between the MSE and SNR for different pilots

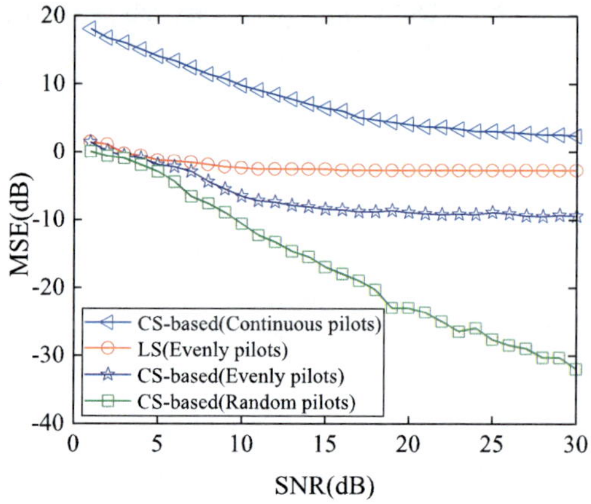

Fig. 2 BER versus SNR for different pilots

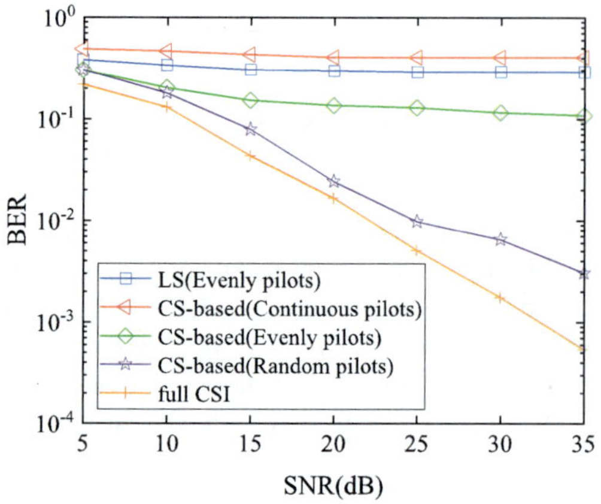

5.2 Mean Square Error Versus BER in Channel Estimation

In order to obtain the proper matrix sensing performance, the pilot's mode position and nonlinear reconstruction algorithm are crucial. In this paper, some degree of performance improvement will be made from OMP, ROMP and CoSaMP reconstruction algorithms. Figures 3 and 4 compare the performance of three compressed estimation algorithms using random-pilot mode, respectively. The performance gap between ROMP or CoSaMP and OMP would be very large if they were not optimized. Therefore, the performance analysis of these three algorithms will be compared using the evenly pilot model and the random search algorithm proposed in this paper, which utilizes the criterion of Eq. (9). The parameter settings of the simulated sparse channel used are shown in Table 2.

The system uses a QPSK coded modulator. The performance evaluation is performed using mean square error (MSE) and signal-to-noise ratio (SNR). The

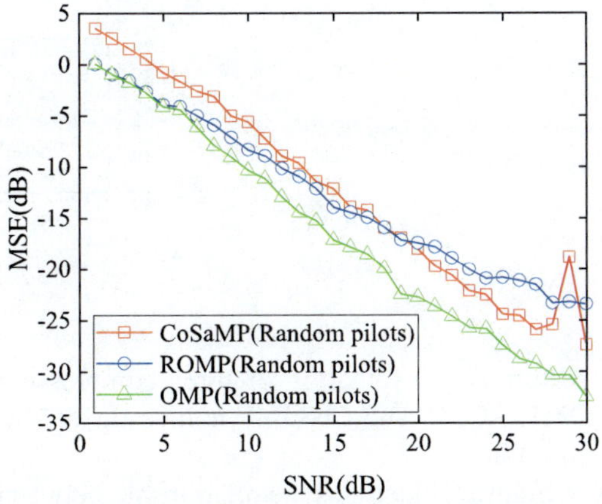

Fig. 3 MSE versus signal-to-noise ratio for three compression estimation algorithms using random-pilot mode

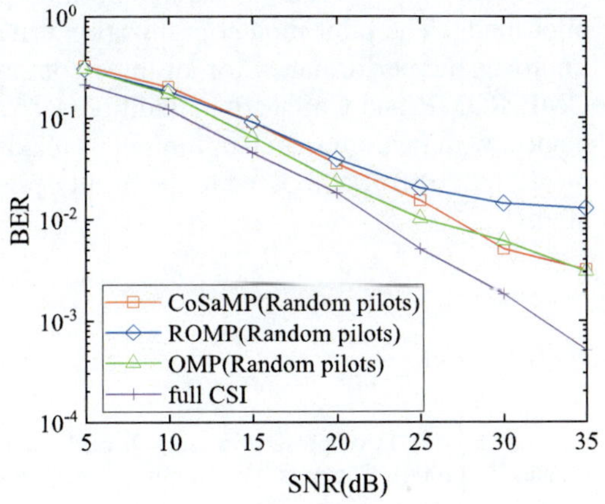

Fig. 4 BER versus SNR for three compression estimation algorithms using random-pilot mode

Table 2 System model simulation parameters for verifying the performance of different estimation algorithms in stochastic pilot mode

Parameter name	Numerical value
Channel length	50
Number of subcarriers	512
Number of pilots	20
Cyclic prefix	128

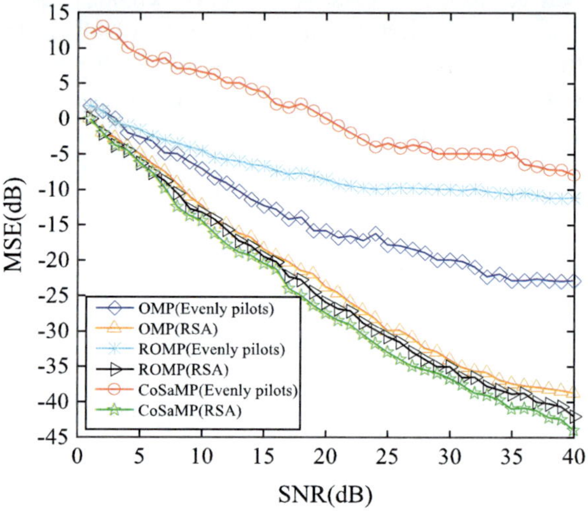

Fig. 5 Comparison of the mean square error performance of OMP, ROMP and CoSaMP compressed-sensing estimation algorithms in the evenly-pilot mode before and after processing by the proposed random search algorithm (RSA)

performance evaluation of three compressed-sensing estimation algorithm types: OMP, ROMP and CoSaMP is investigated by simulation using the random search algorithm.

Figure 5 illustrates the relationship between MSE and SNR for the OMP, ROMP, and CoSaMP compressed-sensing estimation algorithms, using an evenly spaced pilot mode and the random search algorithm proposed in this paper to optimize the pilot mode. The pilot mode optimization criterion of the random search algorithm improves the performance for all three compressed-sensing estimation algorithms: OMP, ROMP, and CoSaMP. In addition, CoSaMP achieves a better gain in performance, with random search optimizing a maximum value of about 35 dB in terms of mean square error compared to the evenly-pilot mode, followed by ROMP at about 30 dB.

References

Baraniuk R (2007) Compressive sensing. IEEE Signal Proc Mag 24(4):118–120, 124

Candes E (2006) Compressive sampling. In: Proceedings of the international congress of mathematicians, Madrid, Spain

Candes E, Romberg J (2006) Quantitative robust uncertainty principles and optimally sparse decompositions. Found Comput Math 6(2):227–254

Candes E, Romberg J, Tao T (2006) Robust uncertainty principles: exact signal reconstruction from highly incomplete frequency information. IEEE Trans Inform Theory 52(2):489–509

Donoho D, Elad M (2003) Optimally sparse representation in general (nonorthogonal) dictionaries via l1 minimization. Proc Natl Acad Sci 100(5):2197–2022

Eladar YC (2009) Uncertainty relations for shift-invariant analog signals. IEEE Trans Inform Theory 55(12):5742–5757

Hwang T, Yang C, Wu G et al (2008) OFDM and its wireless applications: a survey. IEEE Trans Veh Technol 58(4):1673–1694

Tropp J, Gilbert A (2007) Signal recovery from partial information via orthogonal matching pursuit. IEEE Trans Inform Theory 53(12):4655–4666

Shi W, Hao Z, Jie WX et al (2014) Research on channel estimation algorithm in 60 Hz system based on 802.15.3c standard. J Commun 9(1):1–12

Xia P, Zhou S, Giannakis GB (2004) Adaptive MIMO-OFDM based on partial channel state information. IEEE Trans Signal Process 52(1):202–213

A Review of the Development of Artificial Intelligence Electronic Circuit Technology

Zhangchen and Zhangmeng

Abstract Artificial intelligence technology is the new focus of attention of various countries. The development of artificial intelligence technology has put forward new requirements for computing electronic circuits. Deep learning algorithms require massive data training, and traditional computing architectures cannot support the large-scale computing requirements of deep learning algorithms. This paper analyzes different technical routes of artificial intelligence electronic circuits, studies the development trend of artificial intelligence electronic circuit industry in the world and China, analyzes the opportunities and challenges faced by the development of artificial intelligence electronic circuits in China, and provides insights into future artificial intelligence electronic circuits.

1 Introduction

At present, the rapid development of the internet has provided abundant big data resources for artificial intelligence (AI). Thanks to the promotion of data, algorithms and electronic circuits, artificial intelligence technology has risen rapidly. Electronic circuits are the core factor supporting the development of the artificial intelligence industry. Therefore, international traditional IT companies such as Google and Intel have invested a lot of energy in research and development in the field of artificial intelligence electronic circuits, forming different technical routes. On the other hand, the integrated circuit electronic circuit is also a "short board" field in China, which has extremely high strategic significance. Looking back at the development of integrated circuits, it has been driven by technology (Abadi et al. 2016; Bengio et al 2013), architecture and application. With Moore's Law approaching its limits, process improvements have failed to reduce costs. The computationally intensive requirements of artificial intelligence have become one of the main drivers of current electronic circuit technology. The architecture of general-purpose processors has been unable to meet

Zhangchen (✉) · Zhangmeng
Faculty of Information Technology, Beijing University of Technology, Beijing, China
e-mail: Zhangchen132324@163.com; zhangguohui100@petrochina.com.cn

S. Yadav et al. (eds.), *Proceedings of 2nd International Conference on Artificial Intelligence, Robotics, and Communication*, Lecture Notes in Electrical Engineering 1063, https://doi.org/10.1007/978-981-99-4554-2_13

the high demands of artificial intelligence algorithms, and various new architectures have become an important means of improving the performance of current processor electronic circuits. Heterogeneous electronic circuits such as GPU (Graphics Image Processor), FPGA (Field Programmable Gate Array), ASIC (Application-Specific Integrated Circuit) have seized the opportunity one after another, and the emergence of electronic circuits with brain-like neuron structures subverts the traditional structure, bringing new changes to the development of the industry (Benjamin et al. 2014).

2 Definition and Classification of Artificial Intelligence Electronic Circuits

The current artificial intelligence electronic circuit has three meanings: the first refers to the processor electronic circuit that can handle the general tasks of artificial intelligence and has its own core IP (intellectual property); the second refers to the ordinary artificial intelligence algorithm running or embedded The processor electronic circuit; the third type refers to the processor electronic circuit that has the computing efficiency and iterative ability to accelerate one or more tasks such as voice and image (Chen et al. 2017; Dikov et al. 2017). Artificial intelligence electronic circuits can be divided into CPU, GPU, DSP (digital signal processor), FPGA, ASIC and brain-like electronic circuits according to the architecture system (Jouppi 2017). It can be divided into cloud-side and terminal-side electronic circuits according to the usage scenarios, and each side can be divided into training and reasoning according to the task (Fig. 1).

Combining the above two classifications, currently on the cloud side, the main tasks are training, and the electronic circuit architecture is mainly based on CPU, GPU, and FPGA. The end-side is mainly based on the task of reasoning. The end-side electronic circuit cannot bear the huge amount of calculation and has low cost performance. Therefore, the electronic circuit architecture is mainly based on ASIC and DSP.

Fig. 1 Classification of artificial intelligence electronic circuits by task

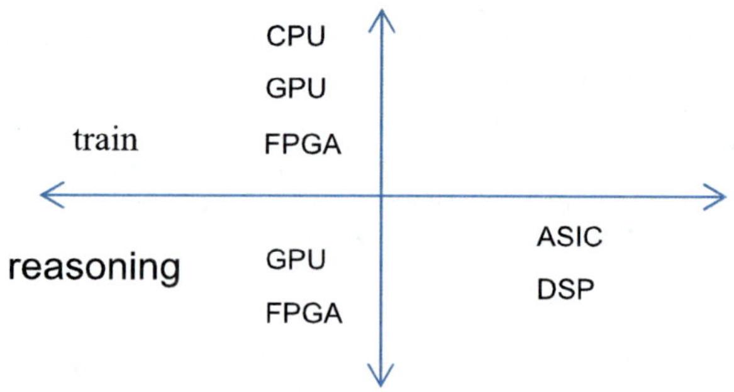

2.1 *Advantages and Disadvantages of Different Electronic Circuits*

Comparing electronic circuits of different architectures with their own advantages and disadvantages in terms of general purpose/specificity, CPU is the most versatile, but has serious latency, high heat dissipation, and the lowest efficiency; GPU is slightly more versatile, faster and more efficient than other electronic circuits, but it is inefficient in the execution stage of neural networks. DSP has fast speed and low energy consumption, but has a single task. At present, mature products are only used as processor IP cores (Lecun et al. 2015); FPGA has the characteristics of low energy consumption, high performance and programmability, and has obvious performance and energy consumption advantages over CPU and GPU. ASICs can be more targeted for hardware-level optimization to achieve better performance. Of course, the design and manufacture of ASIC electronic circuits requires a lot of capital, a long time period and engineering cycle, and the deep learning algorithm is also rapidly iterating. Once ASIC electronic circuits are customized, they cannot be written again. FPGAs have hardware upgradeable. Therefore, at this stage, GPU combined with CPU will be the mainstream of artificial intelligence electronic circuits, and then with the continuous optimization of vision, voice, and deep learning algorithms on FPGA, it will be solidified on ASIC to reduce costs.

2.2 *Technical Paths of Different Electronic Circuits*

The current processor electronic circuits mainly follow two development paths: one is to continue the traditional von Neumann computing architecture (Shi 2016, Sundararajan and Saratchandran 1998), with the main purpose of accelerating hardware computing capabilities, from general-purpose processors (CPUs), graphics processors (GPUs) to digital signals Processor (DSP), then to semi-custom circuit (FPGA) and full-custom circuit (ASIC), the generality of these five types of electronic circuits decreases in turn, which is the direction of upgrading. Another path is to follow the non-von Neumann computing architecture, represented by brain-like electronic circuits, and use the structure of human brain neurons to improve computing power, including the establishment of a complete set of programming environment, compiler and other tools. The current artificial intelligence electronic circuit is evolving along the direction from general-purpose to special-purpose (see Table 1), and in the future, it will move from special-purpose to another level of general-purpose.

Table 1 Artificial intelligence electronic circuit enterprise of the world

Architecture	On behalf of the company	Release time	Introduction	Specificity
CPU	Intel	2017	General-purpose computing processor to accelerate deep learning processing	L1
CPU DSP FPGA TPU	Intel Synopsys Microsoft Google	Apr. 2016 Apr. 2015 Sep. 2016 May 2015	Perform deep learning and neural network tasks Used as processor IP core only Execute Bing's machine learning algorithm Purpose-built integration for deep learning algorithms TensorFlow	L2 L3 L4 L5
TrueNorth	IBM	Oct. 2015	Mimicking the structure of human brain neurons, low power consumption	Subvert classic architecture

2.3 *Industry Situation Analysis of Different Technology Routes*

From an industry perspective, companies with different technical routes have different characteristics.

(1) Companies based on the CPU technology camp are typically represented by Intel Corporation. Intel's advantages have always been in the manufacturing and integration process, and the disadvantage lies in the general architecture design of the CPU, which limits its operating efficiency. Although the current CPU computing in the field of machine learning is greatly reduced, it will not be completely replaced. Intel has launched a series of Xeon processor products for deep learning algorithms. Enterprises take into account both cost and performance factors in the actual use process, and choose CPU so the CPU still plays a big role.
(2) The GPU-based technology camp is typically represented by NVIDIA and AMD. NVIDIA has launched a new Volta architecture GPU for deep learning algorithms, the NVIDIA TeslaV100 GPU computing card. GPUs are primarily engaged in massively parallel computing, run faster than CPUs, and are less expensive than other specialized AI processor electronic circuits. AMD also introduced the world's first 7 nm-class GPU, focusing on GPUs used in mobile devices. Compared with CPU and GPU, GPU can only complete part of the functions of CPU, but the execution speed is much faster. GPUs are more cost-effective for certain applications like crypto currency mining, but CPUs have a broader consumer base.
(3) The DSP-based technology camp is represented by Synopsys and Cadence. Accelerator design based on DSP, such as Synopsys' EV processor, Cadence's

Vision, etc. At present, the design based on DSP has certain limitations. Generally, it is a processor IP core electronic circuit for image and computer vision, which is faster and less expensive.

(4) Representative companies in the FPGA-based technology camp include Xilinx and Altera. FPGA has three major advantages: low unit energy consumption ratio, flexible hardware configuration, and adjustable architecture. However, the use of FPGA has a certain threshold, requiring users to have hardware knowledge. At present, Xilinx and Altera use the latest CMOS node technology to manufacture FPGA electronic circuits, using advanced technology to improve performance.

(5) The ASIC-based technology camp is typically represented by Google. Google launched the tensor processing unit TPU3.0, which is a dedicated logic circuit and is used with the Tensor Flow framework. It is currently dedicated to Google and is not a market-oriented product. In addition, the current algorithm architecture is not completely stable. If the mainstream deep learning algorithm changes greatly, the ASIC electronic circuit cannot change the architecture as quickly as the FPGA and adapt to the changes, which is more expensive for enterprises.

(6) Subvert the classic von Neumann architecture route, typically represented by IBM. In 2016, IBM announced the detailed development plan of electronic circuit, it is a computer electronic circuit based on human brain neuromorphic mixed signals. IBM in 2016 described the architecture, evaluation board family, reference system, and software ecosystem of electronic circuits, with computing units as neurons, memory units as synapses, and transmission units as axons. The technology used in the True North chip is Samsung's 28 nm low-power technology, with 4,096 synaptic cores, each of which contains 256 neurons and 64 KB memory synapses, with a total of 1 million neurons and 2.56 million synapses, the real-time operation power consumption is only 70 MW. But the electronic circuit has the potential to achieve a generalized path in the field of artificial intelligence, but in the short term, it is still far from large-scale commercial production.

3 The Latest Development of Electronic Circuit Technology in the World

At present, due to the characteristics of artificial intelligence application scenario customization and intellectual property protection, Internet giants and traditional IT giants have increased investment in self-developed electronic circuits. The four giants of the Internet—Google, Apple, Facebook, Amazon, and domestic Internet companies such as Baidu and Ali have also launched electronic circuit businesses. For example, Google released TPU3.0, which improved its performance by more than 8 times, and cooperated with the open source framework Tensor Flow to create a closed-loop ecosystem; Microsoft released Project Brainwave, an FPGA-based low-latency deep learning cloud platform; Amazon customized artificial intelligence electronic

Table 2 The situation of artificial intelligence ecological construction of international typical enterprises

On behalf of the company	Architecture	Hardware	Representative products	Application
Google	TensorFlow	TPU	AlphaGo	Mobile application
IBM Microsoft Facebook Amazon	SystemML ML.NET PyTorch XNET	TrueNorth FPGA In research Inferential	Waston Chatbot Echo	Medical service Chatbot Smart speaker
Apple	Turi create	Apple neural engine	Siri	Mobile application

circuits for future applications of Echo device. Baidu released Kunlun electronic circuits for AI applications, Alibaba established Pingtouge Semiconductor Co., Ltd. to focus on customized AI electronic circuits for Alibaba's business scenarios, and Huawei released end-side electronic circuits Kirin 980 and cloud-side electronic circuits Sheng Teng series. At present, American companies have a clear dominant position in the field of AI electronic circuits. Various companies have also built their own hardware and open source platforms to compete for the right to compete in the ecosystem (see Table 2 for details). In the future, the industrial ecosystem that dominates electronic circuits may undergo transformation and upgrading. AI giants like Google and Amazon have reorganized their ecosystems and used cloud services to squeeze the strategic layout of underlying hardware suppliers. Microsoft's Brainwave platform and Facebook's Py-Torch 1.0 software and hardware both compete with Google, and both hope to compete with Google's Tensor Flow and TPU. In addition, ARM released the first-generation processor "Trillium" for AI and machine learning; Nvidia released the new Turing architecture. Artificial intelligence electronic circuits have become the new focus of international industrial competition.

4 Industrial Development in the Field of Artificial Intelligence Electronic Circuits in China

At present, due to the characteristics of artificial intelligence application scenario customization and intellectual From the perspective of China's development, China's technical foundation in the field of integrated circuits is relatively weak, but it started early in the academic research on artificial intelligence electronic circuits. Achieved innovative progress. In the 2016 International Computer Architecture Annual Conference, about 1/6 of the papers cited the Cambrian to carry out research on neural network processors. Not only do start-ups invest enough funds in research, but companies such as Huawei and Baidu also participate in the construction layout and actively grab positions. On the one hand, the AI electronic circuit market in the face of vertical segmentation has broad prospects. With the increasing number of market segments for artificial intelligence application scenarios, the performance of electronic circuits specially customized for some application scenarios is better than that of general electronic circuits, and the terminal electronic circuits are fragmented and diversified, and there is no market monopoly yet., Chinese companies still have more opportunities. The layout of China's artificial intelligence electronic circuit enterprises is shown in Table 2.

However, China faces many challenges as well as opportunities. First of all, the leading companies in the cloud market are located in foreign countries. There is a huge gap between China's cloud electronic circuits and World's technologies. The foreign cloud market technology and ecological construction are mature and have great advantages. There are few companies focusing on cloud electronic circuits in China, and they have not yet formed ecological influence. On the other hand, different enterprises in our country are chasing hot spots quickly, the foundation is weak, and the follow-up is weak. For example, there are 45 start-up companies engaged in artificial intelligence development processors in China, but they are basically engaged in the integrated research and development of voice and visual electronic circuits, with a lot of overlapping positioning, and China has not yet formed an influential electronic circuit-platform-application ecology.

5 Conclusions

The next 10 years will be a critical period for the development and breakthrough of the artificial intelligence industry, as well as an important period for the development of artificial intelligence electronic circuit technology. At this stage, the demand for computing power of artificial intelligence applications is reflected in two aspects. A t first, deep learning algorithms include a large number of computing requirements such as convolution, residual network, and full connection. When Moore's Law is close to the physical limit, process performance improvement will upgrade computing power. Under the premise of decreasing cost performance, the evolution based only on process nodes can no longer meet the needs of rapid growth in computing power; Secondly, deep learning needs to process massive data samples, emphasizing the high parallel computing capability of electronic circuits, and a large number of data handling operations mean Due to the high requirements for memory access bandwidth, the power consumption of memory read and write operations, especially read and write access to off-chip memory, is much greater than the power consumption of computing. Electronic circuits are critical. Therefore, on the one hand, from a technical point of view, the improvement of electronic circuit architecture will become the main means to improve the performance of electronic circuits. From the perspective of products of various enterprises, it is also the main means to iterate the performance of electronic circuits by upgrading different architectures. On the other hand, improving the high-speed communication requirements of computing units and storage units will also become an important trend to improve performance. From the application point of view, terminal electronic circuits have various forms, such as security cameras, smart speakers, smart robots, smart phones, etc. This type of task requires a small amount of calculation, but requires high real-time performance, and pays attention to the energy consumption, heat dissipation, unit energy consumption ratio and other indicators. From an ecological point of view, software and hardware collaborative optimization has become the main way for enterprises to improve their technical capabilities. Pure data and algorithm optimization can no longer meet the needs of enterprises, and enterprises need to use electronic circuits combined with algorithm models to optimize and iterate.

References

Abadi M, Agarwal A, Barham P et al (2016) TensorFlow: large-scale machine learning on heterogeneous distributed systems. arXiv:1603.04467

Bengio Y, Courville A, Vincent P (2013) Representation learning: a review and new perspectives. IEEE Trans Patter Anal Mach Intell 35(8):1798–1828

Benjamin BV, Gao P, Mcquinn E et al (2014) Neuro grid: a mixed-analog-digital multichip system for large-scale neural simulations. Proc IEEE 102(5):699–716

Chen YH, Krishna T, Emer JS et al (2017) Eyeriss: an energy-efficient reconfigurable accelerator for deep convolutional neural networks. IEEE J Solid-State Circuits 52(1):127–138

Dikov G, Firouzi M, Rohrbein F et al (2017) Spiking cooperative stereo-matching at 2 ms latency with neuromorphic hard-ware. In: Biomimetic and biohybrid systems. Springer, Berlin, pp 119–137

Jouppi NP (2017) In-datacenter performance analysis of a tensor processing unit

Lecun Y, Bengio Y, Hinton G (2015) Deep learning. Nature 5:436–444

Shi YX (2016) Research on artificial intelligence process chip technology. Netw Technol 12:11–13

Sundararajan N, Saratchandran P (1998) Parallel architectures for artificial networks: paradigms and implementations. ACM Press, New York, pp 23–27

Stock's Closing Price Prediction Based on GRU Neural Network

Xingyue Yang, Yu Cao, and Xu Cheng

Abstract The closing price of stocks shows different development trends due to different international economic, political conditions, and national policies, which leads to the instability of stock data, and traditional mechanism models cannot make accurate predictions of current environmental changes in a timely manner. Therefore, a study of stock closing prices is required. This paper uses historical closing price data of Maotai as input and through simulation experiments uses five deep learning models to predict and compare the fitting results, and the existing research, such as Convolutional Neural Network-Long Short-Term Memory (CNN-LSTM) model, Long Short-Term Memory (LSTM) model, Gate Recurrent Unit (GRU) model, Convolutional Neural Network-Gate Recurrent Unit (CNN-GRU) model, Convolutional Neural Network-Bi-directional Long Short-Term Memory-Attention Mechanism (CNN-BiLSTM-Attention). The stock price data is used for future rolling prediction. The GRU model is more suitable than the other four models, due to the experiments.

Abbreviations

BiLSTM	Bi-directional Long Short-Term Memory
CNN	Convolutional Neural Networks
CNN-LSTM	Convolutional Neural Network-Long Short-Term Memory
CNN-GRU	Convolutional Neural Network-Gate Recurrent Unit

X. Yang · Y. Cao (✉)
School of Information and Control Engineering, Liao Ning Petrochemical University, Fushun Liaoning, China
e-mail: caoyu@lnpu.edu.cn

X. Cheng
College of Economics and Management, Shenyang Agricultural University, Shenyang, Liaoning, China
e-mail: xucheng@syau.edu.cn

S. Yadav et al. (eds.), *Proceedings of 2nd International Conference on Artificial Intelligence, Robotics, and Communication*, Lecture Notes in Electrical Engineering 1063, https://doi.org/10.1007/978-981-99-4554-2_14

CNN-BiLSTM-Attention	Convolutional Neural Network-Bi-directional Long Short-Term Memory-Attention Mechanism
GA_CART-Adaboost	Combination of CART decision tree optimized by genetic algorithm (GA) and integrated learning Adaboost
GRU	Gate Recurrent Unit
LSTM	Long Short-Term Memory
RNN	Recurrent neural networks

1 Introduction

In 2022, Russia declared war on Ukraine, the stock market has been turbulent, Chinese stocks fluctuated, and technology stocks had been cut in half, causing panic to the social economy and social stability of countries around the world. Therefore, more accurate prediction of stock market data is of great significance for grasping the economic development situation in time, maintaining social stability, and creating wealth.

In order to scientifically predict the development trend of stocks, researchers have proposed many stock prediction models, such as Combination of CART decision tree optimized by genetic algorithm (GA) and integrated learning Adaboost (GA_CART-Adaboost), (Xiaotong and Wengang 2021) Long Short-Term Memory (LSTM), (Jie 2020) Convolutional Neural Networks (CNN) (Hinton and Salakhutdinov 2006). Among them, Long Short-Term Memory (LSTM) is considered to be the most suitable deep learning in the financial field (Ruyi 2021). On this basis, Zhang Wei proposed to analyze stock market volatility based on text mining investor sentiment (Wei 2015). Text analysis can effectively reflect stock market sentiment, but it is difficult to quantify specific stock price changes. Shen Shanshan put forward a Convolutional Neural Network combined Long Short-Term Memory (CNN-LSTM) short-term stock price prediction founded on attention mechanism (Shanshan and Qiumin 2022). The method first applies a convolutional neural network to obtain useful feature components from data convolutions, and then uses LSTM to predict the sequence of extracted feature components. The output vital vector of the essential hidden layer at each time node is multiplied by the corresponding weight and summed, and then the larger weight is assigned to the important feature component as the final feature expression of the model. Meng Yi used a stock prediction focused on CNN-BiLSTM and attention mechanism (Convolutional Neural Network-Bi-directional Long Short-Term with Memory-Attention Mechanism) (Yi and Qingjuan 2021). Additionally, the built model's prediction results were compared to those of CNN, LSTM, Bi-directional Long Short-Term Memory (BiLSTM), CNN-LSTM, and CNN-BiLSTM in an empirical analysis. It was discovered that CNN-BiLSTM combined attention, the force mechanism, performed better in predicting stock data with multiple features and long time series. A multi-factor quantitative stock selection method based mostly on Gate Recurrent Unit (GRU) neural networks was proposed by Ouyang Mingzhe (Mingzhe 2020). The first step of this strategy is to establish a multi-factor stock selection index system, classify it into five types of factors, and then screen the indicators. The second step is to establish a GRU neural network stock sequence classification prediction model, and compare its prediction effect with the Logistic and RNN models. The third step is to test the GRU multi-factor stock selection strategy. In the process of GRU neural network stock sequence classification and prediction, new stock selection factors are synthesized and reverse tests are applied to test the significance of the strategy. This method proposes a complete set of stock trading strategies, which can effectively test the feasibility of the model, but there are too many input indicators for the GRU model, and it is impossible to quantitatively analyze the importance of each indicator. Cheng Mengfei proposed a stock price prediction model by multi-scale called TLEMD-LSTM-MA (TELM). The TELM model first makes the closing price to break down into components on multiple time scales using empirical modes, then uses different prediction methods according to the different oscillation frequencies of the components, and finally takes the sum of the predicted values of all components as the final closing value (Mengfei and Shuping 2022). This model can improve the prediction accuracy of stock closing prices, but it requires too much data processing in the early stage. Feiyan et al. (2021) used an LSTM neural network to predict short-term price trends, and it was found that training an LSTM model with weekly data was more accurate than stock predictions trained with daily data. Kexin (2020) proposed the research on stock trend prediction of deep neural network. First, the stock text information is classified based on Naive Bayes, and then the network is constructed by BiLSTM and CLSTM. Finally, the final fully connected layer outputs the prediction result. It is found that the established model can to a certain extent give accurate and effective prediction of stock trends. Changwei (2020) proposed a stock prediction research based on the deep bidirectional LSTM-RNN model. Adam is improved and the improved algorithm is applied to the prediction model of BiLSTM, and it is found that the BiLSTM model's prediction is reasonable, accurate, and feasible.

To be able to find the model that predicts a single stock price most accurately, the LSTM model and CNN-LSTM model, the GRU model and CNN-GRU model, and the CNN-BiLSTM with Attention model are all examined in this work. For comparison and tuning, use Moutai's closing price data from January 5, 2015 to April 25, 2022 for training fitting prediction. The results picture that the GRU model's prediction effect is the best. Then, the GRU model is designed to predict the movement of Maotai stock price 60 opening days after April 25, 2022. At present, there are many stock prediction models, which retain most of the information of the corresponding stock for a single stock forecast, which is reliable and practical. Therefore, the models compared in this paper are of great significance for predicting stock prices.

2 Data Introduction

This article is based on NetEase Finance's daily data on Moutai stocks from January 5, 2015 to April 25, 2022. There are 1777 daily data, the total training set is the Moutai's closing price data from January 5, 2015 to end of 2021, and the test set is the data from January 4, 2022 to April 25, 2022. On this basis, remove the closing date, and continuously predict the closing price of Moutai for 60 opening days.

2.1 Data Preprocessing

The input of unprocessed data to the neural network will produce large errors, so the experiment normalizes the input historical closing price data. The normalization formula is as follows:

$$x_{\text{std}} = \frac{x - x_{\min}}{x_{\max} - x_{\min}} \tag{1}$$

x_{std} is normalized data, x is the current closing price data, $x_{\min}$ is the lowest closing price data, $x_{\max}$ is the lowest closing price data.

3 Principles of Stock Forecasting Models

According to the previous work introduced in the introduction, it is found that the deep learning algorithm with significant effect on stock prediction has LSTM (Ruyi 2021), CNN-LSTM (Chenyang 2021), CNN-BiLSTM-Attention (Yi and Qingjuan 2021), GRU (Mingzhe 2020), CNN-GRU (Weijie et al. 2021). Therefore, this paper mainly selects these five algorithms for comparison.

3.1 LSTM Model

In this paper, the normalized data of the first 9 closing prices is used as input, and the results are obtained after training with LSTM, and then the results are reverse-normalized to obtain the output. It uses a specialized RNN called the LSTM, which is primarily developed to address gradient disappearance and gradient explosion issues during long-sequence training. It performs better than a standard RNN during long-sequence training. Figure 1 shows the LSTM's organizational chart.

The input gate is represented by i_t, the forget gate is represented by f_t, o_t is output gate. The forget gate phase is mainly used to forget the input selectively from the previous node. Choose to remember what is essential, forget what is not. The input

Fig.1 LSTM model

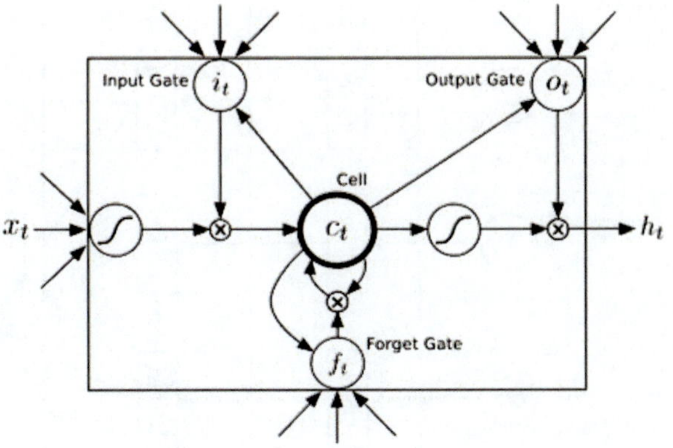

gate does not always "remembers" the input stage. The output gate determines the current state. Each gate control equation is as follows:

$$i_t = \sigma\left(W_i \cdot \left[h_{t-1}, x_t\right] + b_i\right) \tag{2}$$

$$f_t = \sigma(W_f \cdot \left[h_{t-1}, x_t\right] + b_f) \tag{3}$$

$$\tilde{C}_t = \tanh(W_C \cdot \left[h_{t-1}, x_t\right] + b_f) \tag{4}$$

$$o_t = \sigma(W_O\left[h_{t-1}, x_t\right] + b_o) \tag{5}$$

$$C_t = f_t * C_{t-1} + i_t * \tilde{C_T} \tag{6}$$

$$h_t = o_t * \tanh(C_t) \tag{7}$$

3.2 CNN-LSTM Model

The CNN-LSTM combined model first takes the normalized data of the first nine closing prices as input, and then uses CNN to convolve the data sequence to extract its feature components. Then utilize the LSTM to predict the extracted feature components to get the final vector v. Then normalize to get the output value *y*. The CNN-LSTM model is shown in Fig. 2.

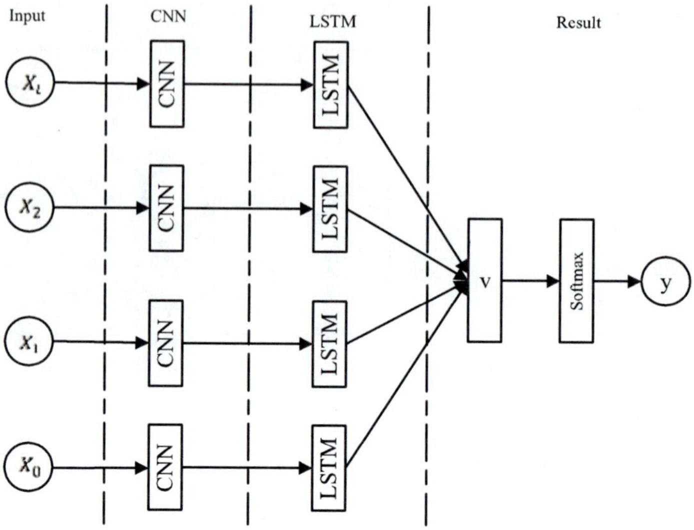

Fig. 2 CNN-LSTM model

3.3 CNN-BiLSTM with Attention Model

When predicting the stock price, the CNN-BiLSTM with Attention model first convolves to simplify the data features, then uses the bidirectional LSTM to train the upper and lower data information, and finally uses the attention mechanism to get the final vital vector, which is normalized to obtain the output value y (Mengfei and Shuping 2022).

CNN-BiLSTM with attention's model diagram is pictured in Fig. 3.

3.4 GRU Model

GRU functions almost the same as LSTM. It mainly combines one forget and one input gate into one update gate, mixes hidden state and "cell state", with some other changes. Figure 4 shows the GRU's structure.

The respective gating equations are as follows:

$$z_t = \sigma(W_z \cdot [h_{t-1}, x_t]) \tag{8}$$

$$r_t = \sigma(W_r \cdot [h_{t-1}, x_t]) \tag{9}$$

$$\tilde{h_t} = \mathrm{Tan}h(W \cdot [r_t * h_{t-1}, x_t]) \tag{10}$$

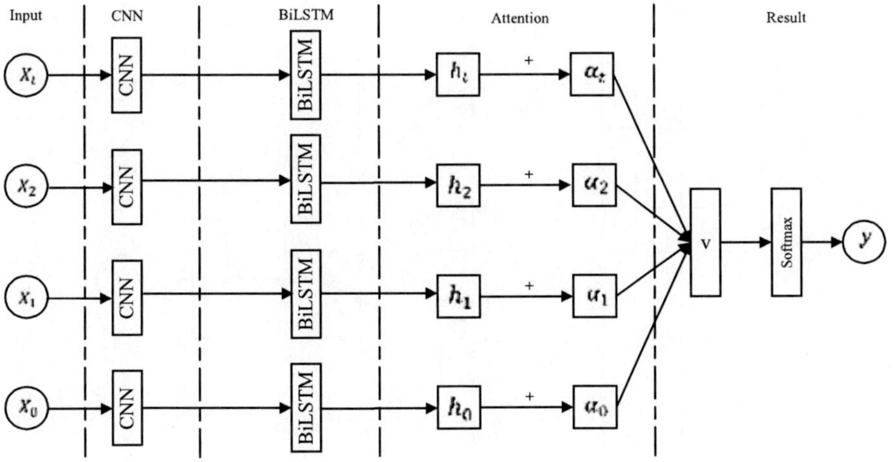

Fig. 3 CNN-BiLSTM-attention model

Fig. 4 GRU model

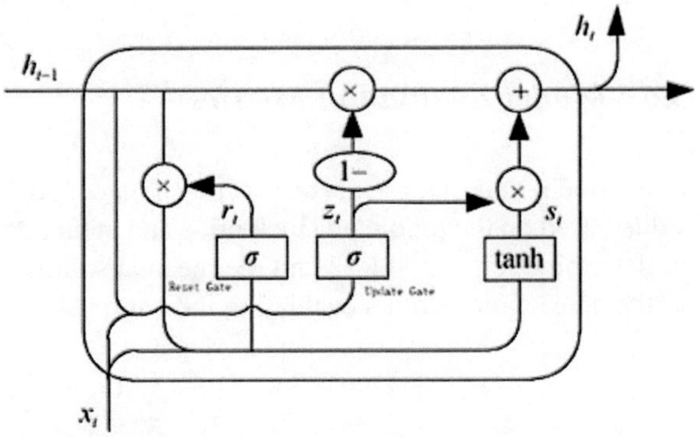

$$h_t = (1 - z_t) * h_{t-1} + z_t * \tilde{h}_t \tag{11}$$

3.5 CNN-GRU Model

The CNN-GRU combined structure diagram is shown in Fig. 5.

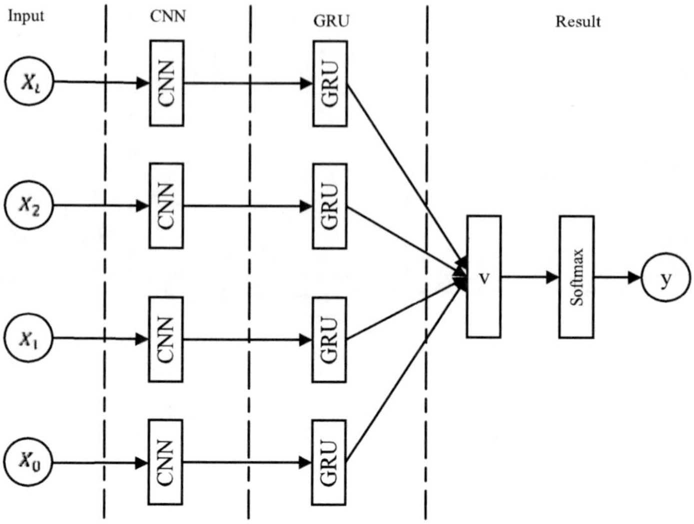

Fig. 5 CNN-GRU model

3.6 Model Evaluation Metrics

After model training, use the test set to validate model predictions, calculate the error value based on the predicted stock price and actual data. In this paper, the coefficient of determination called R^2 and the mean absolute error called MAE are employed as the prediction model's evaluation indicators. The calculation formula is as below

$$R^2 = 1 - \frac{\sum_i (\hat{y}_i - y_i)^2}{\sum_i (y_i - \overline{y})^2} \tag{12}$$

$$MAE = \frac{1}{n}\sum_{i=1}^{n} |y_i - \hat{y}_i| \tag{13}$$

$\hat{y}_i$ is the test set sample' predicted value. y_i is the test set samples' true value. $\overline{\overline{y}}$ is the mean of the test set samples. n is the test set samples' total number. When MAE is smaller, R^2 closer to 1 means the better the model.

4 Analysis of Results

In this experiment, the LSTM, GRU, CNN_LSTM, CNN_GRU, CNN_LSTM_ Attention models were used to predict the daily closing price of Moutai stocks. The data range of the training set was the closing price data of Moutai from January

5, 2015 to December 31, 2021. The data range of the set is the closing price data of Moutai from January 4, 2022 to April 25. According to the experimental results, the forecasting situation of Moutai stock price of each model is analyzed. Then, the prediction curves of the five models are compared, and the prediction performance of each model is evaluated under the same error index. Finally, according to the experimental comparison results, the most suitable model for Maotai's stock price is obtained.

4.1 Model Prediction

LSTM Model

Input the training samples into the LSTM model. The model's network layer is LSTM, one layer of linear layer, the time step is surely set to 6 steps, the hidden layer neurons' number is hidden size = 30, the loss function only uses MSE, and the Loss of the training set is. The lowest value can converge to 9e−5. The LSTM prediction effect of Maotai stock price is shown in Fig. 6.

According to the index parameters in Fig. 6 and Table 1, the fitting effect and prediction effect of the model are close to the true value.

CNN-LSTM Model

The network layers contributed by this model are one convolutional layer, two LSTM layers, and one linear layer. The time step sets 6 steps, the convolution kernel is 1*1, the needed activation function of the convolution layer and LSTM is RELU, and the model optimizer is "Adam". After 100 times of training, the model loss can converge to 4e−4 at least. Figure 7 shows the prediction effect of stock price with CNN-LSTM model.

Fig. 6 LSTM model Moutai stock price predict

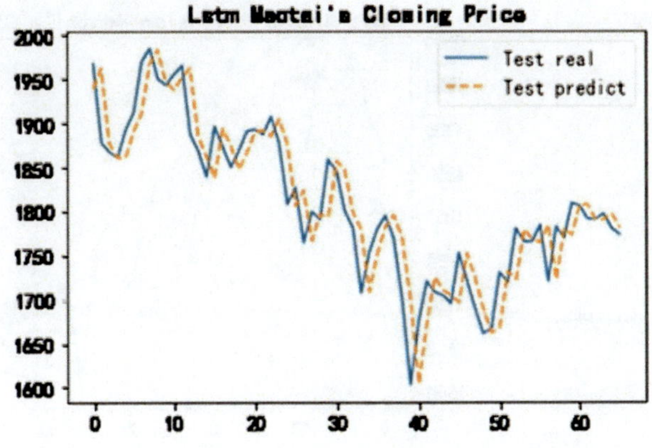

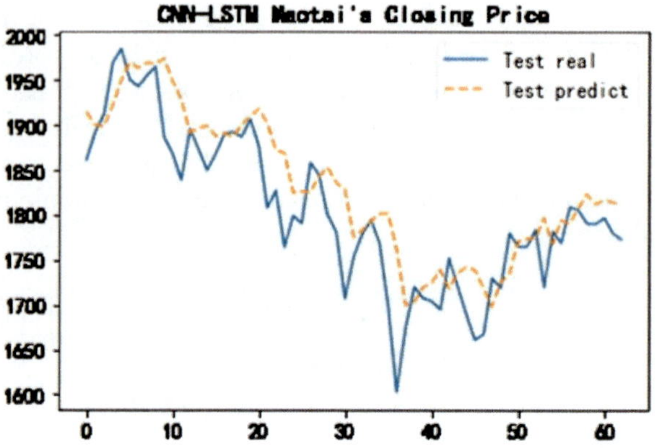

Fig. 7 CNN-LSTM model Moutai stock price predict

According to the CNN-LSTM model's prediction total effect and the error indicators in Table 1, the CNN-LSTM model's prediction total effect is worse than the LSTM model.

CNN-BiLSTM-Attention Model

The model contains a convolutional layer, a bidirectional LSTM layer, a linear layer, and an Attention layer. The set time step is 6 steps, the convolutional layer and the bidirectional LSTM layer activation function are SIGMOID and TANH, the activation function of the Attention layer is SIGMOID, and the model optimizer is "Adam". The Loss of the training set can converge to a minimum of 4.35e−04. The prediction of CNN-BiLSTM with Attention is shown in Fig. 8.

According to the CNN-BILSTM-Attention prediction diagram and Table 1, the CNN-BiLSTM with Attention model is less effective than two of the other models LSTM related.

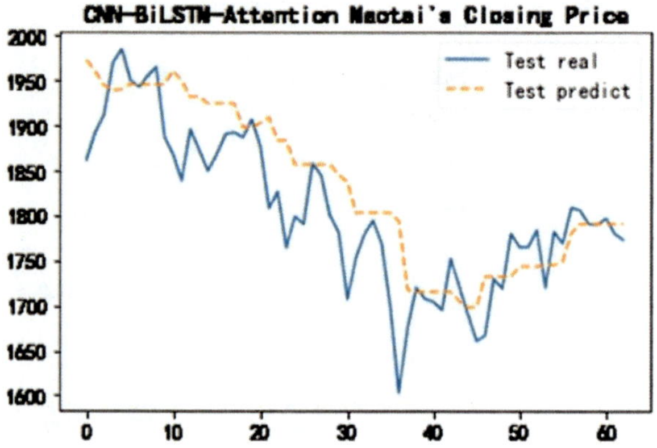

Fig. 8 CNN-BiLSTM-attention model Moutai stock price predict

Fig. 9 GRU model Moutai stock price predict

GRU Model

Input the data into the GRU model, which consists of one GRU layer and one linear layer. The activation function of GRU is TANH, the feedback activation function is HARD_SIGMOID, the time step is 9 steps, and the optimizer is "Rmsprop". LOSS converges to 8e−5. The GRU model fitting and prediction results are shown in Fig. 9.

According to the GRU model fitting and prediction graph, and the prediction error parameter indicators in Table 1, it can be obtained that the GRU model is surely better than the CNN-LSTM and CNN-BiLSTM-Attention. The prediction accuracy is roughly in line with the LSTM model.

CNN-GRU Model

The model is mainly made up of one convolutional layer, one pooling layer, and two GRU layers. The vital activation function of the simple convolutional layer is RELU, the total number of hidden layer neurons is hidden size = 40, the used activation function of GRU is TANH, and the optimizer is "Adam", using MSE as the loss function, the lowest Loss value of the training set can converge to 1e-4. The fitting effect and prediction effect of the CNN-GRU model of Moutai's stock price are shown in Fig. 10.

According to the error indicators in Figs. 6, 7, 8, 9, 10 and Table 1, when opposed to the other four relevant models, the GRU can more accurately fit and predict stock prices, which verifies the superiority of this model.

4.2 GRU Forecast for the Next 60 Closing Price Chart

The forecast chart for rolling forecast of future trading days using GRU neural network is shown in Fig. 11.

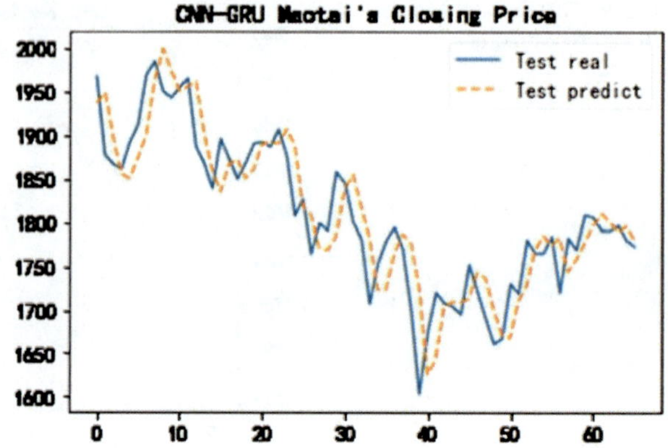

Fig. 10 CNN-GRU model Moutai stock price predict

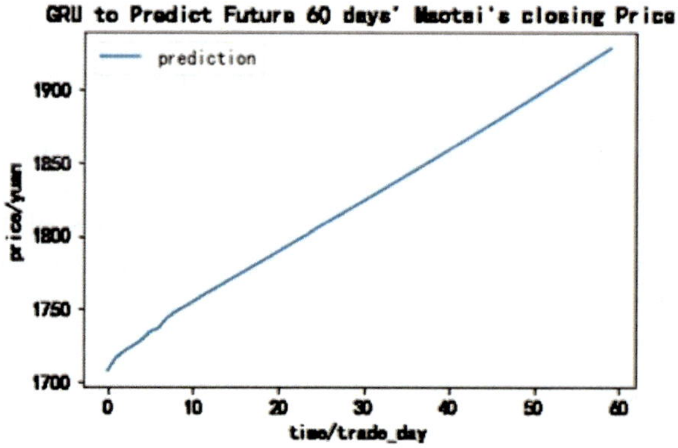

Fig. 11 GRU model Moutai stock price 60 days predict

4.3 Prediction Error Comparison

See Table 1.

Table 1 Different models of Maotai stock price prediction error indicators

Model	R^2	MAE
LSTM	0.7808	32.195
CNN-LSTM	0.7215	34.115
CNN-BiLSTM-attention	0.7084	39.136
GRU	0.8036	28.999
CNN-GRU	0.7439	36.221

5 Concluding Remarks

Based on the daily data of Moutai stock price, this paper establishes the LSTM model, CNN-LSTM model, CNN-BiLSTM-Attention model, GRU model, and CNN-GRU model to predict Moutai stock price data. It is proved by experiments that the GRU network model based on Moutai stock price data has higher prediction accuracy than the other four models. Therefore, the GRU model can surely better predict the development trend of Moutai's stock price, which is of great significance for analyzing the stock's daily closing price trend of a single stock.

Consider current environmental factors for stock price forecasts. Therefore, in the future, this study will consider combining news, investor sentiment, and stock earnings reports and other factors to be quantified and input as model features; it will also consider improving the GRU model to improve the prediction accuracy (Qibin et al. 2020).

Acknowledgements The author welcomes the insightful correction and criticism of anonymous reviewers and editors.

Funding The authors gratefully thank the truly great support as follows: 1. the National Natural Science Foundation of China (Grant 71902121), 2. Scientific Research Project of Educational Department of Liaoning Province (Grant L2020031), and 3. Key Research and Development Project of Liaoning Province (Grant 2020JH2/10300040).

Ethics Declarations

Ethical approval and participation consent This paper is totally original and has never been submitted nor published elsewhere.

Consent for Publication Does not apply.

Availability of Data and Materials Correspondent authors can provide data sets analyzed during the current study period upon reasonable requests.

Competing Interests No competitive interest.

Contributions CY designed the study and greatly helped in the final vital draft of the manuscript. YXY wrote the manuscript writing, contributed to the experiment and analyzed the data and wrote some experimental codes. CC made insightful revisions of the manuscript.

References

Changwei L (2020) Research on stock prediction based on deep bidirectional LSTM-RNN model. Heilongjiang Univ. https://doi.org/10.27123/d.cnki.ghlju.2020.001270

Chenyang L (2021) Research on stock price prediction and quantitative stock selection based on CNN-LSTM. Northwestern Univ. https://doi.org/10.27405/d.cnki.gxbdu.2021.001928

Feiyan D, Shaoqi C, Fengqi Z, Jiahui P (2021) Short-term price trend prediction based on LSTM neural network. Comput Syst Appl 30(04):187–192. https://doi.org/10.15888/j.cnki.csa.007855
Hinton GE, Salakhutdinov RR (2006) Reducing the dimensionality of data with neural networks. Science 313:504–507
Jie Z (2020) Empirical analysis of stock prediction based on LSTM. Shandong University
Kexin C (2020) Research and application of stock trend prediction based on deep neural network. Nanjing Univ Posts Telecommun. https://doi.org/10.27251/d.cnki.gnjdc.2020.001022
Mengfei C, Shuping G (2022) Multi-scale stock prediction based on deep transfer learning. Comput Eng Appl 1–13
Mingzhe O (2020) Multi-factor quantitative stock selection strategy based on GRU neural network. Zhongnan Univ Econ Law. https://doi.org/10.27660/d.cnki.gzczu.2020.000129
Qibin W, Fangmin D, Shuifa S (2020) Text classification method based on BiLSTM-attention-CNN hybrid neural network. Comput Appl Softw 37(09):94–98+201
Ruyi Y (2021) Stock prediction analysis based on deep learning. China Collect Econ 24:105–106
Shanshan S, Qiumin L (2022) CNN-LSTM short-term stock price prediction based on attention mechanism. Software 43(02):73–75
Wei Z (2015) Research on investor sentiment and stock market performance based on Weibo text mining. Shandong University
Weijie C, Weihui J, Xibei J (2021) Stock index price prediction based on CNN-GRU joint model. Inform Technol Informatization 09:87–91
Xiaotong L, Wengang C (2021) Research on multi-feature stock prediction model based on GA_CART_Adaboost. China Water Transp (Second Half Month) 21(08):52–53+56
Yi M, Qingjuan X (2021) Stock prediction based on CNN-BiLSTM and attention mechanism. J Nanning Normal Univ (Natural Science Edition) 38(04):70–77. https://doi.org/10.16601/j.cnki.issn2096-7330.2021.04.010

Random Forest Algorithm for Forest Fire Prediction

Gaolun Fan

Abstract Forest fires threaten the national forestry resources and also endanger the safety of residents. In the past, forest fire research has mostly focused on the prediction of forest fires, and understanding the relevant laws governing the occurrence and progression of forest fires is a crucial assurance for conducting scientifically sound fire prevention and suppression operations. Deep learning, which is based on machine learning, is a useful technique for predicting the burned area of a forest fire. The dataset used in this study is sourced from the UCI website, about forest fires in northeastern Portugal, with 517 instances and 13 attributes. The dataset is preprocessed, characterized, and analyzed to derive the pattern of fire occurrence, using the analysis and processing through the historical data of forest fire occurrence and the corresponding meteorological information. The support vector regression model, decision regression tree model, artificial neural network, and random forest regression model were used to establish the forest fire burning area prediction models, and the various model methods were compared and tested. According to the results, the performance of random forest prediction results is lower compared with the other three models. In this experiment, the random forest model predicts the probability of forest fires, which can effectively guide humans to protect against forest fires. It is important for reducing the safety risk of residents' production and life and improving the efficiency of securing large natural resources.

1 Introduction

Forests as the "green lung" of the earth, forest resources have an important impact on the ecological environment of our country, and play a significant role in the protection of the environment in the world. Forests not only keep the atmosphere clean and purify the air effectively, but also maintain water and soil, and play a vital role in improving people's life as well as the living environment. Due to natural

G. Fan (✉)
Electronic Information Engineering, Qiqihar University, Qiqihar 161003, China
e-mail: 164304305@stu.cuz.edu.cn

S. Yadav et al. (eds.), *Proceedings of 2nd International Conference on Artificial Intelligence, Robotics, and Communication*, Lecture Notes in Electrical Engineering 1063, https://doi.org/10.1007/978-981-99-4554-2_15

factors, in drier seasons and areas in the forest, spontaneous fires are easily formed by the high temperature of ground vegetation and fallen leaves. The fires that occur in such cases are major natural disasters, the fires spread quickly and directly cause the destruction of forest resources. Forest fires have always threatened the national forest resources and the safety of residents, the research of forest fire prediction have been the focus of forest fire research. If algorithmic model prediction can be used to find the pattern and development trend of forest fire occurrence, it will provide a scientific and effective guarantee for forest fire prevention and fighting work.

Previous studies on the topic of forest fire prediction have been based on the direction of hardware sensors, using satellites, infrared scanners, and local sensors (Qin 2020; Zhang 2009; Qiu et al. 2020), but this solution is not suitable for application in forest fire prevention and early warning due to the delay in satellite positioning and the cost of equipment as well as the maintenance of sensors. In recent years, several studies have also investigated the use of traditional machine learning models for forest fire prediction (Dai et al. 2020; Li et al. 2016; Bai et al. 2021), and there are also studies on new deep learning neural networks for forest fire prediction (Sun et al. 2019; Zhao 2019). However, few studies have considered both including machine learning, deep learning, and ensemble learning, and have conducted comparative studies of multiple models to select the best model for prediction.

In this experiment, for the research topic of forest fire prediction, this study tried the commonly used traditional machine learning models, deep learning, and ensemble learning to compare different models for forest fire prediction and select the best model for prediction. Finally, the experiments found that random forest had better results for forest fire prediction, indicating that the tree model has advantages for doing the regression task, while the emerging neural network has difficulty in reaching the effectiveness of machine learning and ensemble learning for forest fire prediction.

2 Methods

2.1 Data Preparation

2.1.1 Dataset Introduction

This experiment is based on a dataset of forest fires in northeastern Portugal, with 517 instances and 13 attributes, to derive the pattern of fire occurrence, which is processed by analyzing historical data of forest fire occurrence and the corresponding meteorological data. Using regression models, a forest fire area prediction model was developed, and various modeling methods were compared and tested.

2.1.2 Feature Encoding

Some of the features of the dataset, such as month and day, are character type. Therefore, it needs to choose a suitable encoding method to convert the character type to value type, to facilitate the subsequent feature analysis of the dataset and training. The encoding method chosen for this experiment is LabelEncoder.

2.1.3 Feature Normalization

There are 13 feature attributes inside the dataset, and the value range of each feature attribute is not in the same scale, which is not conducive to modeling and computation. So, all 13 feature attributes are normalized so that the value range is in the same scale, which is convenient for model training. The formula of the data normalization can be found in Eq. (1).

$$Y = (X - X_{\min})/(X_{\max} - X_{\min}) \tag{1}$$

2.1.4 Correlation Principle

The Pearson correlation coefficient was considered in this paper for evaluating the relationship between variables.

$$r = \frac{\operatorname{cov}(x, y)}{\delta_x * \delta_y} = \frac{\sum_{i=1}^{n}(x_i - \overline{x})(y_i - \overline{y})}{\sqrt{\sum_{i=1}^{n}(x_i - \overline{x})^2}\sqrt{\sum_{i=1}^{n}(y_i - \overline{y})^2}} \tag{2}$$

2.1.5 Correlation Analysis and Visualization

The Pearson correlation coefficient ranking can be found in Fig. 1.

2.1.6 Model Evaluation Metrics

(1) Mean Squared Error

$$\text{MSE} = \frac{1}{n}\sum_{i=1}^{n}\left(X_{i\_\text{real}} - X_{i\_\text{pred}}\right)^2 \tag{3}$$

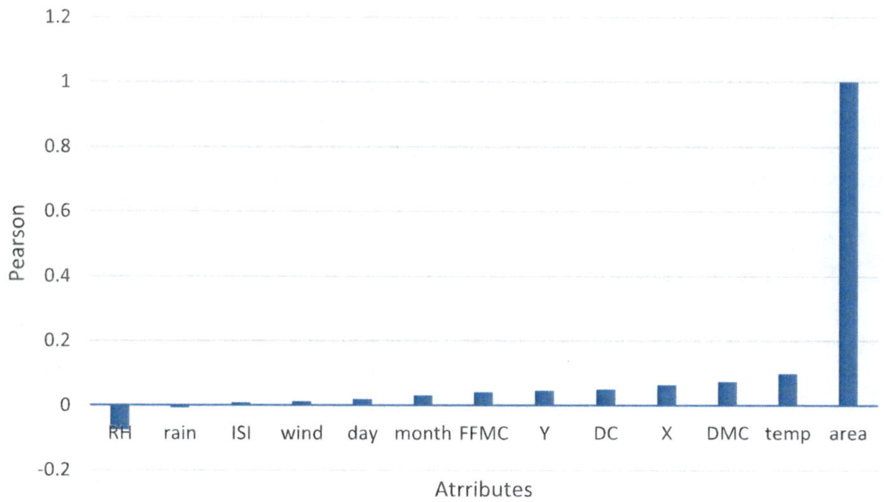

Fig. 1 Pearson correlation coefficient ranking chart

(2) Mean Absolute Error

$$\mathrm{MAE} = \frac{1}{n}\sum_{i=1}^{n}\left|Y_{i\_\mathrm{real}} - Y_{i\_\mathrm{pred}}\right| \tag{4}$$

2.2 *Method Introduction*

2.2.1 Support Vector Regression

SVR, a crucial application subset of support vector machines, seeks to identify a regression plane where all of a set's data are most closely related. It has a threshold ε near the regression line compared to the general linear regression, which makes the generalization ability of SVR enhanced. There are some characteristics of the kernel function: (1) The nonlinear transformation function's shape and parameters may not be known; (2) Avoiding "dimensional catastrophe" to greatly reduce the amount of computation.

The form and parameters of the kernel function change as an implicit change in the mapping from the input space to the feature space, thus having an impact on the nature of the feature space, and finally this study chose the appropriate kernel function by comparing different kernel functions.

$$f(x) = \sum_{i=1}^{m} (\hat{a}_i - a_i) k(x_i^T x) + b \quad (5)$$

2.2.2 Artificial Neural Networks

Artificial neural networks are similar to neuronal networks in the human brain, and there are different ways of network connections. The strong performance has been demonstrated in many studies (Yu 2019; Sae-Tang 2022). Each neural node is an output function, also called an activation function. Different weighting values exist for the connections between two neural nodes, and the output results of the final neural network vary depending on the topology of the connections, the weighting values, and the activation function.

2.2.3 Decision Regression Tree

The ideal division point is discovered, and the sample set is divided into the root node. The optimal division point (j, s) is used to split the data into sub-nodes, the attribute value less than s is divided into one node, and the attribute value greater than s is divided into another node. This division point is a specific value (s) of an attribute (j). The value of the node output is the mean value at this node, which can be used to check whether the termination condition is met. If it is, the cycle comes to an end; if not, it is divided according to one of the many termination conditions, which can include the maximum decision tree depth or the minimum number of samples of child.

2.2.4 Random Forest Regression

The principle of random forest is to average multiple decision trees on the basis of decision trees, which makes the resulting model more robust since each tree is subject to high variance.

3 Results and Discussion

3.1 Regression Performance of Different Models

After the data pre-processing, and the model training times are adjusted to the optimal, the following are the prediction results of the four models. The performance of several models can be found in the following Figs. 2, 3, 4, 5 and 6.

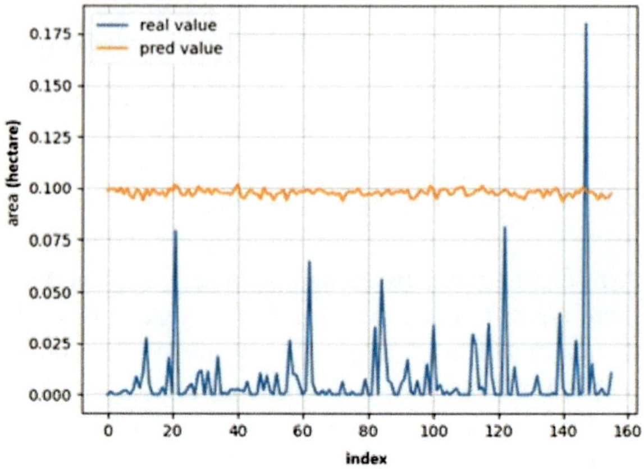

Fig. 2 Support vector regression

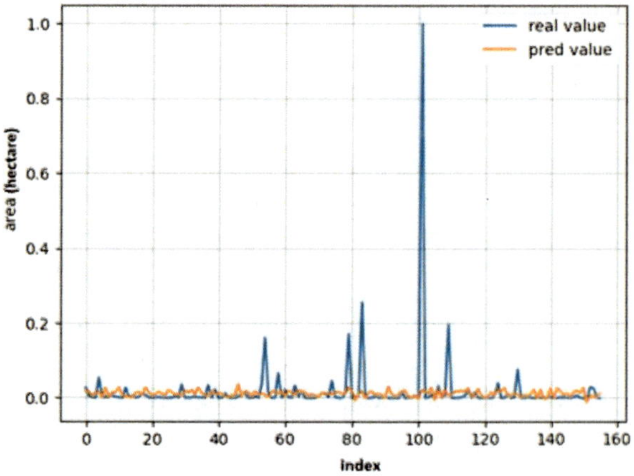

Fig. 3 Neural networks

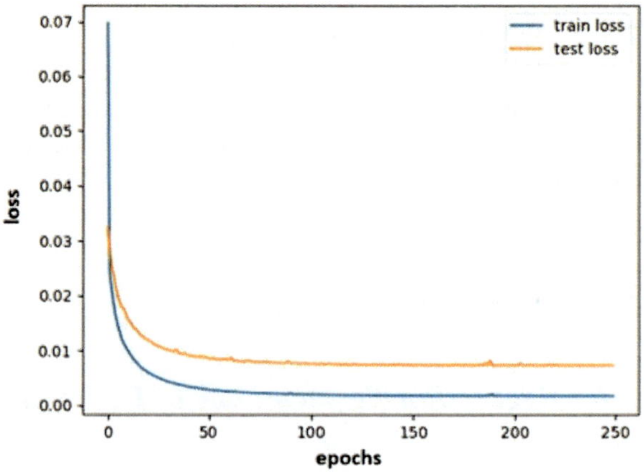

Fig. 4 Loss function comparison

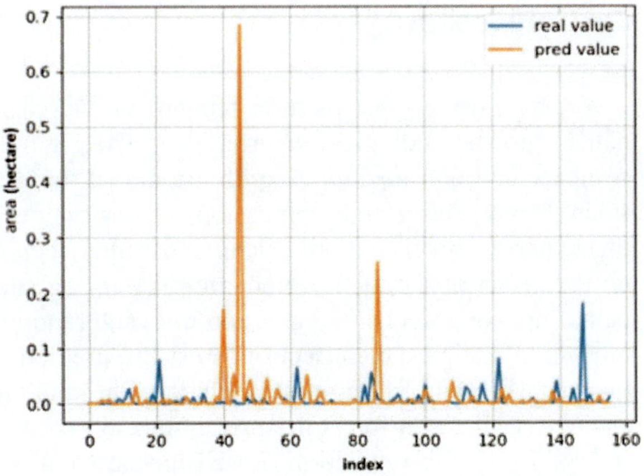

Fig. 5 Decision regression tree

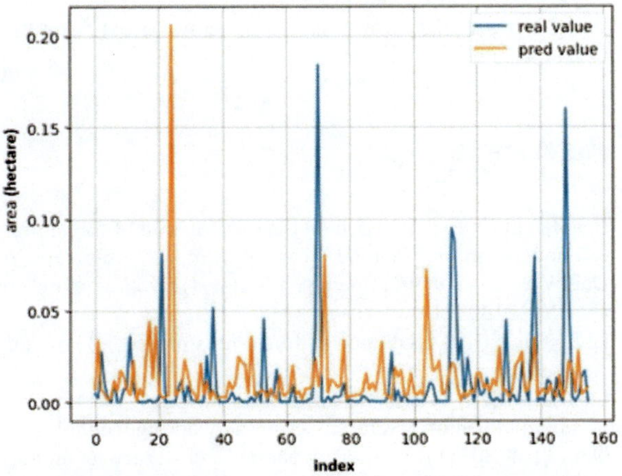

Fig. 6 Random forest regression

According to Table 1, regression performance of Decision Tree is better than Support Vector Regression and Neural Network, and Random Forest Regression is the best.

Table 1 Comparison of evaluation indicators of each model

	MSE	MAE
SVR	0.008531518	0.091442632
Neural network	0.007295903	0.023570115
Decision regression tree	0.004053667	0.01637432
Random forest regression	0.001008551	0.015887487

4 Conclusion

Forest fires bring great harm to human survival and life while causing irreparable damage to the ecological environment. The significance of this experiment is to compare and optimize an algorithmic model that helps predict forest fires for the scientific prevention and suppression of forestry fires. Based on the existing research, this experiment covers both machine learning, deep learning and ensemble learning on the effect of forest fire prediction, by trying support vector regression model, neural network, decision regression tree, and random forest. This study established models to compare the effect of forest fire prediction, and based on the evaluation indexes MSE and MAE, it is concluded that random forest has smaller error value and better effect on forest fire prediction, so random forest is more suitable for the task of forest fire prediction. Based on the solutions adopted in the current stage of research on forest fire prediction, it is necessary to continuously optimize the accuracy of forest fire prediction while trying to save costs, and it is hoped that there will be more efficient ways to prevent forest fires in the future.

References

Bai H et al (2021) Research on the construction of forest fire prediction model based on Bayesian model averaging method–a case study of Dali Prefecture. Yunnan Prov J Beijing for Univ 43(5):9

Dai W et al (2020) A method for forest fire prediction analysis based on machine learning:, CN112132321A

Li E et al (2016) Prediction of forest fires based on least squares support vector machine. Hans J Data Min

Qin YL (2020) Forest fire monitoring and early warning system based on NB-IOT narrowband communication and multi-sensor networking technology. Internet Things Technol 6:14–19

Qiu J et al (2020) A method, apparatus and storage medium for forest fire prediction based on soil moisture CN112016744A

Sae-Tang W (2022) A hybrid automatic defect detection method for Thai woven fabrics using CNNs combined with an ANN. In: IWAIT, vol 12177. SPIE, pp 119–124

Sun L et al (2019) Forest fire prediction method based on deep learning of meteorological factors. For Sci Technol Dev 4(3)

Yu Q (2019) Semantic segmentation of intracranial hemorrhages in head CT scans. In: 2019 ICSESS. IEEE, pp 112–115

Zhang J (2009) Forest fire detection system based on wireless sensor network. In: 2009 4th IEEE conference on industrial electronics and applications. IEEE, pp 520–523

Zhao L (2019) Research on forest fire smoke detection algorithm based on convolutional neural network. China New Technol New Prod 12

Underwater Image Clearing Algorithm Based on the Laplacian Edge Detection Operator

Xingzhen Li, Siquan Yu, Haitao Gu, Yuanyuan Tan, and Lin Xing

Abstract When shooting underwater, limited by different camera hardware and lighting, underwater images are generally of low quality and blurred details. Moreover, due to the different absorption rates of the R, G, and B three-color wavelengths, the color of underwater photos is distorted. This affects the feasibility of studying underwater images. The majority of currently available techniques for improving underwater photographs mainly solve the problem of overall denoising and brightness enhancement of underwater images while ignoring the edge details of the image. As a solution to the aforementioned issues, we suggest a method using a double extraction network structure model based on the Laplacian operator, which can dehaze underwater images more quickly. One is to use downsampling and upsampling to reduce the excessive distortion of the image during training, and the other is to use the Laplacian operator to reduce the blurring of the edge to enhance the clarity of the edge and use the blue–green model for multiscale fusion. We adopted MSE loss and proved its superiority through experiments and index analysis by training images of different underwater scenes.

X. Li · S. Yu · H. Gu (✉) · L. Xing
State Key Laboratory of Robotics, Shenyang Institute of Automation, Chinese Academy of Sciences, Shenyang 11000, Liaoning, China
e-mail: ght@sia.cn; 1172869363@qq.com

Institutes for Robotics and Intelligent Manufacturing, Chinese Academy of Sciences, Shenyang 11000, Liaoning, China

X. Li · Y. Tan
School of Artificial Intelligence, Shenyang University of Technology, Shenyang 110870, Liaoning, China

S. Yadav et al. (eds.), *Proceedings of 2nd International Conference on Artificial Intelligence, Robotics, and Communication*, Lecture Notes in Electrical Engineering 1063, https://doi.org/10.1007/978-981-99-4554-2_16

1 Introduction

Underwater picture enhancement is a technique used to improve low-quality underwater photographs, which is a challenging task. Because underwater scenes are different from land, they have complex water characteristics and high concentrations of particles in suspension, and the quality of the collected images is seriously degraded for the reason that light is being absorbed and scattered. Additionally, when underwater robots and autonomous underwater vehicles are operating, such as submarine cable inspection, undersea docking, geological exploration of a submarine area, and other effective decisions, enhancement technology is needed.

At present, traditional picture enlargement algorithms are split into two distinct sections: enhancement algorithms as the basis for physical models and nonphysical models. The former requires mathematical modeling for the underwater image construction process to estimate model parameter information to figure out what went wrong with the whole picture so that we may fix the underwater photo where it was before it was ruined. The most classical physical model enhancement algorithm is He et al. (2010) using the dark channel prior removal of single image haze (DCP) algorithm to calculate the pixel points and found that it is very effective to find the darkest point locally for uniform fog removal and can also use the fog concentration to estimate the distance of objects. However, the application of this algorithm is limited, which leads to the overall darkening of the image. To this end, Zhu et al. (2015) developed a linear model to represent the image depth with the haze, and subsequently, with the use of supervised learning, the model's settings were fine-tuned to produce a satisfactory depth map. Then, the model of the atmosphere is employed to reconstruct the fogless image so that the restored image has good color saturation and natural color. Similar to the dark channel previous technique, Chao and Wang (2010) employed local pixel blocks of dry photographs to estimate the depth of murky water. The other is a nonphysical model, which mainly uses the mathematical characteristics and statistical characteristics of the image itself for restoration, such as histogram equalization (Pizer et al. 1987). This algorithm uses the contrast of the part of the image with a concentrated gray value to enhance through nonlinear transformation and the contrast of the part with a sparse gray value to weaken so that to accomplish defogging, an image's general grey value dispersion is uniform. The retinex defogging algorithm (Fan et al. 2019) uses the principle of object color constancy and the Gaussian function to enhance images. Wavelet and homomorphic filtering algorithms (Wen et al. 2016) use different scales to decompose the original image signal locally and then obtain the low-frequency and high-frequency characteristics of the image. When defogging and enhancing the image, it is necessary to suppress the low-frequency signal part of the image and relatively strengthen the high-frequency signal of the image to achieve defogging and enhancement. Although traditional methods are simple and efficient, most of them have limited application scope, poor portability, and robustness.

With the continuous improvement of computer technology, AI (Jiang et al. 2022) is applied in many aspects, such as voice recognition, image processing, intelligent robots, autonomous vehicles, and financial technology. At present, image enhancement algorithms based on depth learning mainly have two research directions: first, on the basis of the scattering model, clear images can be gathered from the original model by estimating or restoring some of its parameters through a neural network; second, clear images can be obtained directly by employing an end-to-end CNN neural network model with fogged photos as the input and defogged images as the result. The DeHazeNet (Cai et al. 2016) algorithm applies the model of air scattering on the hazy picture provided as input for image defogging and enhancement. GCANet (Chen et al. 2019) is based on the end-to-end defogging enhancement algorithm. It uses unsupervised learning to conduct model training, that is, to directly establish the connection between fuzzy images and clear images. It uses smooth hole convolution to replace general hole convolution, which not only increases receptive fields but also enables adjacent pixels to contain common information. Naik et al. (2101) recommended a simplified neural network structure, Shallow-UWnet, which retains performance while demanding a few variables. Liu et al. (2019) used a cyclic consistent adversarial network (CycleGAN) in order to provide the convolutional neural network model with training data in the form of synthetic underwater photographs, introduced super-resolution to reconstruct the model, and proposed an underwater enhancement model with a residual architecture, because underwater AUVs rely on various sensors for work and intelligent decision-making, suspended particles and color distortion in the water and other factors will influence the caliber of visual information. Fabbri et al. (2018) used the method of generating a countermeasure network (GAN) to enhance the quality of underwater visuals, in an effort to enhance the visual input of autonomous action.

At present, whether traditional methods or depth-based learning methods are used to achieve the clarity of underwater images, there are the following problems: most networks use single structure networks to extract image features and rarely consider the edge clarity of image information. To solve these potential problems, this paper proposes a dual feature extraction network structure model algorithm based on the Laplace edge detection operator. First, we use a dual structure and multiple local features to retrieve the characteristics of the input image so that more image details can be captured. To reduce the blurring or even loss of edge information, the Laplace operator is employed to get rid of the border around a picture.

This paper mostly consists of the following sections: An overview of relevant study of underwater image clearing and analysis of the existing problems is described in Sect. 2; and Sect. 3 introduces the network structure and loss function of the proposed underwater image enhancement method; Comprehensive experiments are conducted in Sect. 4 to test the validity of the suggested technique; and Sect. 5 summarizes the full text and gives possible future research issues.

2 Related Work

A few decades ago, most of the edge detection operators proposed by people have been applied to target recognition and interpretation of images, but few have been applied to image clarity. Over time, technique utilizing deep learning can be employed to improve edge detection, while classical techniques still have their place.

The traditional edge detection method calculates the differential or second-order differential of pixels in the picture to identify the edge pixel. (1) The most typical edge detection operators are the Roberts operator (Ma et al. 2008), Prewitt operator (Torre and Poggio 2008), Sobel operator (Kittler 1983), and the widely used Canny operator (Canny 1986). (2) Structured image edge detection operator (Hou et al. 2007) combined with (Canny 1986) solved the response time relationship between pixel edges and adjacent edges; Similarly, (Wang et al. 2017) for the low accuracy of feature vector solution time and normalized notch detection in image edge, wavelet was used to extract edge features and construct multiscale structure for sampling. The two-layer active contour model (Liang et al. 2015) is used to extract the contour from top to bottom to capture the weak edge of the image. Using the structured learning framework of the random decision forest (Dollár and Zitnick 2013), the problem of predicting the local edge mask is formulated.

With the ongoing advancement of deep learning (Shi et al. 2021) technology, particularly the appearance of CNN (Albawi et al. 2017), the ability to sample images has been greatly improved, and it is more accurate and delicate. Using CNNs to detect edges has become a new trend. Xie and Zhuowen (2015) proposed the whole nested edge detection (HED) method. Subsequently, Poma et al. (2020) suggested a novel HED (holistically nested edge detection)-based edge detection technique and Xception networks, which can be utilized without training or fine-tuning for any edge detection job. Using a complete convolutional encoder decoder network, Yang et al. (2016) were able to recognize contours in image, laying a foundation for the application of full convolution in contour detection. Technique for edge identification of multiscale movement items was suggested by Wang et al. (2017).

3 Our Model Design Method

Following the Laplacian edge detection operator's dual extraction network concept, we present an approach for imaging images taken below the water's surface, including an extracting features component, a dehazing module, and an edge sharpening module, as shown in Fig. 1.

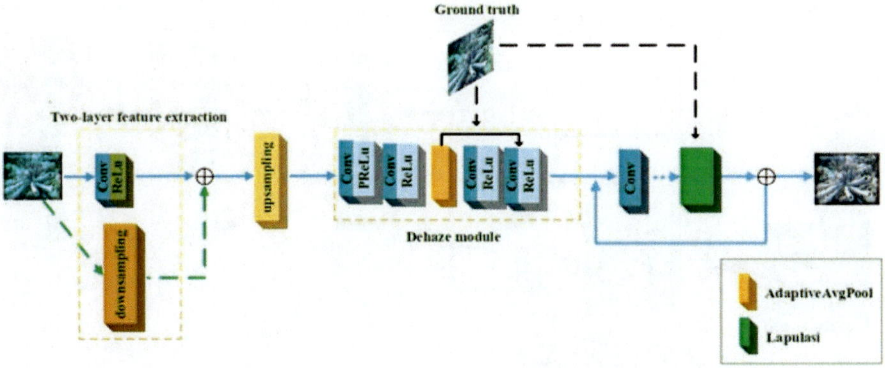

Fig. 1 Network architecture

3.1 Feature Extraction Module

Due to the inherent blur and distortion of underwater images, if conventional convolution (LeCun et al. 1989) is used to extract information and features of underwater images, it will cause more or less loss of image details. Therefore, to reduce the loss of information during feature extraction, the model structure we adopted is the upsampling (Gupta 2012) structure and conventional convolution for double-layer extraction. Information loss can be kept to a minimum by always filling in the picture with 0, which makes feature maps and the output data image and prevents the picture from shrinking over time due to convolution or causing the edges of the image to disappear. Due to factors including variation in lighting conditions and color distortion, underwater photographs are seldom of uniform quality, if all attention is given to the whole, it will reduce the quality of the images after training. Therefore, multiple parts of the images are extracted and captured to prevent unnecessary omissions. First, the underwater image is routinely extracted for image features, and then we use downsampling to extract and capture multiple parts of the image. Figure 2 demonstrates that, according to the size of the image, convolution feature extraction is performed by downsampling. The color image has 3 channels, and its convolution kernel is 3*3. Then, the activation function LeakyReLU is added with a value of 0.1. The activation function LeakyReLU is introduced, mainly during backpropagation, and during the training process, the gradient can also be calculated for the part of the input less than 0, which can avoid the disappearance of the gradient and the aliasing problem in the gradient direction in time.

Due to the effect of downsampling, the input image becomes a thumbnail and then enters upsampling using reverse convolution, still using 0 padding, and its convolution kernel is 3*3. Because we need to perform pixel-level classification on the image during the processing of images, we need to restore the feature map to the original image through upsampling after convolution to extract features. BatchNorm2d (Ioffe and Szegedy 2015) is added after the inversion convolution layer for data normalization because the activation function added to the deep neural network

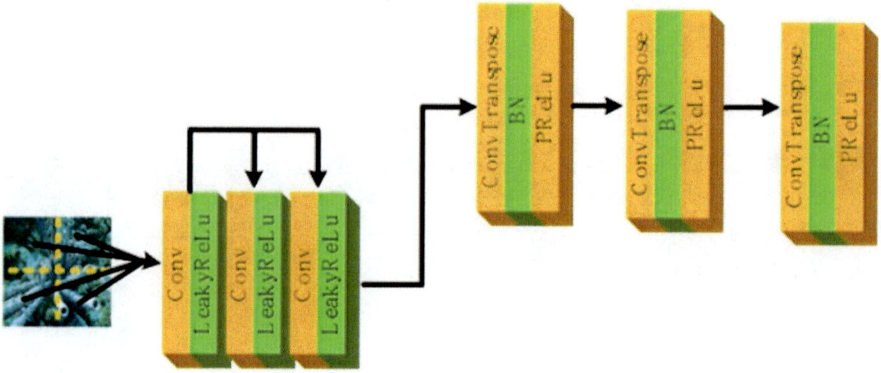

Fig. 2 Multiple local sampling

will gradually follow the continuous training of the neural network, the deepening of the number of network layers before the nonlinear transformation. Its gradient distribution will gradually shift or change, resulting in the disappearance or explosion of the underlying neural network gradient during backpropagation, which in turn makes the convergence rate slower. Adding BN pulls the input value of the neural network during training back to a certain standardization method. The mean value is 0 and the variance is 1 in the standard normal distribution interval, so that the quantity of the activation intake is in the range where the nonlinear functionality has a significant impact on the output, so that increase in gradient size prevents gradient from vanishing. Greatly speed up training. Then, below is the BN formula:

$$\text{Batch mean: } \mu_\beta = \frac{1}{m}\sum_{i=1}^{m} x_i \tag{1}$$

$$\text{Batch variance: } \sigma_\beta^2 = \frac{1}{m}\sum_{i=1}^{m} \left(x_i - \mu_\beta\right)^2 \tag{2}$$

$$\text{Batch normalization: } \hat{x}_i = \frac{x_i - \mu_\beta}{\sqrt{\sigma_\beta^2 + \varepsilon}} \tag{3}$$

$$\text{Displacement: } y_i = \gamma \hat{x}_i + \beta \tag{4}$$

where m denotes the batch size; x_i represents the input; y_i refers to the output; γ and β are the learning parameters.

The input image data undergo 3 downsamplings and 3 upsamplings. However, in the training process, it will be restored to the original image, which provides a better reference for subsequent dehazing enhancement so that fewer visual details are destroyed, and the original image is restored to the greatest extent.

3.2 Defogging Module

This dehazing module includes 4 sets of convolution kernels with a size of 3*3. The adaptive average pooling layer can dynamically generate the size of the pooling kernel, adopt the principle of "template maximization," round up to obtain the size of the pooling kernel, calculate the equidistant position points on the facet map, and utilize the "minimum principle" to round down when calculating the starting coordinates. According to the set output signal size (output-size) and the size of the input signal, the width of the typical pooling layer and the moving step are automatically calculated. Zero-padding is performed to prevent image edge data from being lost after back-and-forth learning and training, and it can also increase the receptive field (Isola et al. 2017). The network can make better use of available pictures and better estimate data characteristics if the receptive field is increased in size. The activation function uses PReLU with parameters. In the negative range, the slope of PReLU (Wang et al. 2018) is small, which can avoid the dead ReLU problem. Another point is that although PReLU is a linear operation in the negative range, although the slope is small, it does not tend to trend. at 0. The formula of PReLU is shown in (5).

$$\text{PReLU}(x_i) = \begin{cases} x_i & \text{if } x_i > 0 \\ a_i x_i & \text{if } x_i \leq 0 \end{cases} \tag{5}$$

where x_i is the input value of the nonlinear activation function in the ith channel in the characteristic map of the neural network and a_i is the learning arguments of the activation function.

The initial segment of an input picture is used to reconstruct the not clear image, and the second part of the network estimates the picture background, the target image and the label image for comparative learning to achieve an ideal dehazing effect.

3.3 Edge Sharpening Module

Due to the inferior quality of the original underwater photo, the loss of details is relatively significant, especially the blurring or even removal of specifics about the image's edges. For this reason, we employ an identifying edge operator, which is a Laplacian template operator, as shown in formula (6). Local grayscale differences in images are computed using the Laplacian operator (Tai et al. 2008), and it is an image neighbourhood enhancement algorithm derived through second-order differentiation. It can be found from the template that when neighbouring pixels have the identical grayscale value, the template's convolution function yields a result of 0; a favourable outcome from the template convolution procedure is obtained when the grayscale of the middle pixel is greater than the grayscale of the surrounding pixels on average. When the grayscale of the centrcenter pixel is lower than the typical grayscale of other pixels in the vicinity, the template undergoes deleterious convolution. It is the job of the convolution kernel, an operation, to execute a convolution on the value data produced by the defogging module.

$$\begin{pmatrix} 1 & 1 & 1 \\ 1 & 8 & 1 \\ 1 & 1 & 1 \end{pmatrix} \begin{pmatrix} 1 & 1 & 1 \\ 1 & -8 & 1 \\ 1 & 1 & 1 \end{pmatrix} \tag{6}$$

Use (6) as the convolution kernel, a function that performs a convolution on the defogging module's output value information, and use the Laplacian template to perform the label image convolution procedure to complete the required edge information as much as possible.

Because the center value of the Laplacian template has positive and negative values, there is a slight difference for extraction, as shown in Fig. 3, and we use the Laplacian template with a negative center.

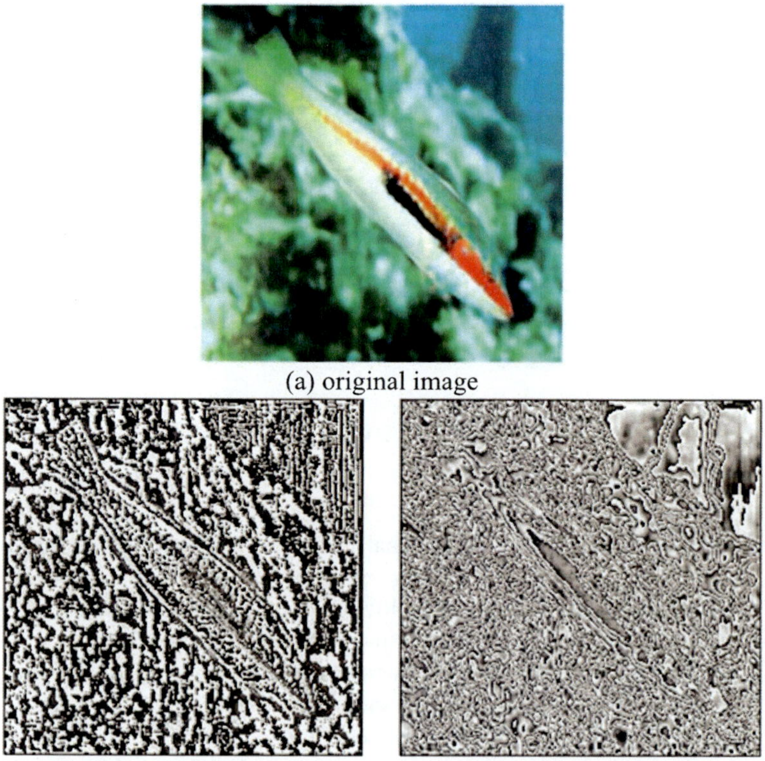

(a) original image

(b) center negative value

(c) center positive value

Fig. 3 Laplacian template edge detection

3.4 Loss

To generate a dehazing image to avoid being too smooth and lacking the authenticity of the photo in the visual sense, we use an MSE (mean squared error), that is, the MSE, which can be regarded as a kind of L2 loss, which is another popular regression loss functions. MSE is the squared summation of variance between what was expected and what was received. Since the functionality curve is flat, consistent, and can be computed in all locations, the gradient descent process can be employed with relative ease, making it a popular loss function. Convergence is facilitated by the fact that the gradient lowers as the error goes down, and this is valid even if the learning rate remains constant; hence, the minimal value can be reached more swiftly.

$$L_{\mathrm{MSE}}(x_i, x_i') = \frac{\sum_{i=1}^{n} (x_i - x_i')}{n} \tag{7}$$

where x_i is the actual value and x_i' is the predictive value.

4 Experimental Analysis

4.1 Experimental Environment Parameter Configuration and Dataset Selection

The hardware configuration of this experiment is Intel® Xeon(R) Gold 5222 CPU @ 3.80 GHz × 16, 64-bit processor, NVIDIA Quadro RTX 5000 GPU for computing acceleration, the operating system is 64-bit Ubuntu18.04, using the PyTorch1.11.0 framework, and Python3.7 is used for network training and implementation.

The epoch of the underwater image dehazing enhancement iterative training is 150 times, and the fixed learning rate is mainly used, which is 0.001. As a result of inaccuracy brought on by fluctuating learning rates, the batch size is 1, and the optimizer used is Adam because the update of Adam parameters is not affected by the stretching transformation of the gradient, which is suitable for problems with sparse gradients or large gradient noises and is also suitable for scenarios with large-scale data and parameters.

The EUVP (Enhanced Underwater Visual Perception) dataset of the Minnesota Interactive Robot and Vision Laboratory contains different sample sets of paired and unmatched photos of low visual quality, which is conducive to promoting the supervised training of underwater image enhancement models. Training datasets are paired datasets. The dataset contains 2185 images of different water depths, such as underwater geology, fish, and underwater plants. The size of each image is 248*248.

4.2 Image Evaluation Norm

Subjective evaluation will be scored according to each person's own wishes, and there is often too much deviation. The objective evaluation of image quality can accurately and automatically determine the "good or bad" quality of the reconstructed image in a scientific way, which can be used to replace the human visual system to see and recognize the image.

For the most part, we rely on both comprehensive and no-reference picture quality evaluations.

(1) Full reference image quality evaluation

a. PSNR

PSNR refers to the signal-to-noise ratio at the apex. Usually, after picture reconstruction and compression, the output image will appear different from the label. Assessing the final image's appearance after processing, the PSNR value is used to quantify whether a certain processing program makes people work hard, so the lower the distortion, the higher the PSNR value. Below is the PSNR formula.

$$\text{PSNR} = 10\log_{10}\frac{255}{\sqrt{\left|\hat{x} - x\right|^2}} \tag{8}$$

where $\hat{x}$ is the output image value, x is the input picture.

b. SSIM (Structural Similarity)

Consistency between photos can be measured using SSIM. Similarity between images is measured on a scale from 0 to 1, with a higher number indicating more similarity.

$$\text{SSIM}(\hat{x}, x) = \frac{(2\mu_x\mu_{\hat{x}} + c_1)(2\sigma_{x\hat{x}} + c_2)}{\left(\mu_x^2 + \mu_{\hat{x}}^2 + c_1\right)\left(\sigma_x^2 + \sigma_{\hat{x}}^2 + c_2\right)} \tag{9}$$

where μ_x and $\mu_{\hat{x}}$ are the gray mean values of the normal input image x and $\hat{x}$ the enhanced, respectively; σ_x and $\sigma_{\hat{x}}$ are the variance, $\sigma_{x\hat{x}}$ are the covariance of x and $\hat{x}$, c_1 and c_2 are small numbers, avoiding the denominator being 0.

(2) No reference image quality evaluation

a. Underwater Color Image Quality Evaluation (UCIQE)

Evaluation of the degree to which color casts, blurring, and poor contrast can be quantified by employing a sequential blending of color, brightness, and darkness. The formula is as follows.

$$\text{UCIQE} = c_1 \times \sigma_c + c_2 \times \text{con}_1 + c_3 \times \mu_s \tag{10}$$

where σ_c represents the standard deviation of the chroma of the target image, con_1 represents average variation in the contrast of the target picture, μ_s represents the typical deviation of the average saturation of the target image, and c_1, c_2, c_3 are the coefficient constants set to 0.4680, 0.2745, and 0.2576, respectively.

b. Underwater image quality metric (UIQM)

Three attributes of underwater picture are considered: underwater image colorimetry (UICM), underwater image intelligibility metric (UISM), and underwater image contrast metric (UIConM). The formula is as follows.

$$\text{UIQM} = c_1 \times \text{UICM} + c_2 \times \text{UISM} + c_3 \times \text{UIConM} \tag{11}$$

where c_1, c_2, c_3 is the coefficient constant set to 0.4680, 0.2745, and 0.2576.

4.3 Analysis of Experimental Results

This paper's recommendations about the network model are beneficial to dehazing and enhancing the brightness of underwater images, as shown in Fig. 4.

There is a positive impact of the network model suggested in this study on dehazing and enhancing the brightness of underwater images, as is seen from the figure.

The experiment's goal is to see how well the proposed network model and approach function in practise. As it pertains to underwater photography, this is mostly about dehazing the blue-green channel and fixing the red channel to restore a single picture. The dark channel prior removes the single image haze (DCP) (He et al. 2010) method, removes the water dispersion (Chao and Wang 2010) method, and the

Fig. 4 Enhancement effect of underwater image dehazing

(a) original image (b) DCP (c) ROWS

(d) Shallow-UWnet (e) OUR

Fig. 5 Enhancement effect of four kinds of underwater foggy images

shallow neural network architecture, namely, Shallow-UWnet (Naik et al. 2101). As shown in Fig. 5, this paper's technique is compared to others.

In addition to the abovementioned subjective visual comparison, three application ratios of PSNR, SSIM, and UIQM are also used.

A wide range of objective metrics are used to compare the overall and structural details of noise and raw images. Table 1 displays superior SSIM, PSNR, and UIQM ratings for the approach presented in this research. Although UCIQE is lower than method (d), the comprehensive value still achieves a relatively ideal effect. Therefore, a strategy indicated in this study can not only greatly enhance the underwater photo dehazing enhancement ability but also suppress excessive noise, enhance brightness, and enhance the picture's general clarity.

Table 1 Comparison of image defogging evaluation indicators

Evaluation norm	b	c	d	e
PSNR	14.018	19.739	17.481	**24.057**
SSIM	0.622	0.759	0.727	**0.857**
UCIQE	0.597	0.600	**2.348**	0.577
UIQM	3.817	3.804	3.331	**4.374**

5 Conclusion

This research proposes a novel imaging technique for improving undersea dehazing. This method is based on the RGB distribution characteristics of undersea photo, and the combination of a deep learning network design and Laplacian operator is used for experimental verification and optimization. By eliminating atmospheric haze, the resulting undersea photo is of superior quality. In the future, we will continue to increase our method's capacity to adapt to a wide variety of situations, especially in deep water and shallow water.

Acknowledgements The funding for this project came from the National Natural Science Foundation of China (Grant No. 62206274).

References

Albawi S, Mohammed TA, Al-Zawi S (2017) Understanding of a convolutional neural network. In: 2017 international conference on engineering and technology (ICET). Ieee

Cai B et al (2016) Dehazenet: an end-to-end system for single image haze removal. IEEE Trans Image Process 25(11):5187–5198

Canny J (1986) A computational approach to edge detection. IEEE Trans Pattern Anal Mach Intell 8(6):679–698

Chao L, Wang M (2010) Removal of water scattering. In: 2010 2nd international conference on computer engineering and technology, vol 2. IEEE

Chen D et al (2019) Gated context aggregation network for image dehazing and deraining. In: 2019 IEEE winter conference on applications of computer vision (WACV). IEEE

Dollár P, Zitnick CL (2013) Structured forests for fast edge detection. In: 2013 IEEE international conference on computer vision, pp 1841–1848https://doi.org/10.1109/ICCV.2013.231

Fabbri C, Islam MJ, Sattar J (2018) Enhancing underwater imagery using generative adversarial networks. In: 2018 IEEE international conference on robotics and automation (ICRA), pp 7159–7165https://doi.org/10.1109/ICRA.2018.8460552

Fan D et al (2019) Scale-adaptive and color-corrected retinex defogging algorithm. In: 2019 3rd international conference on electronic information technology and computer engineering (EITCE). IEEE

Gupta M (2012) Filter design for gray image by down-sample and up-sample. In: Fourth international conference on digital image processing (ICDIP 2012), vol 8334. SPIE

He K, Sun J, Tang X (2010) Single image haze removal using dark channel prior. IEEE Trans Pattern Anal Mach Intell 33(12):2341–2353

Hou J, J-H Ye, Li S-S (2007) Application of Canny combining and wavelet transform in the bound of step-structure edge detection. In: 2007 international conference on wavelet analysis and pattern recognition, vol 4. IEEE

Ioffe S, Szegedy C (2015) Batch normalization: accelerating deep network training by reducing internal covariate shift. In: International conference on machine learning. PMLR

Isola P, Zhu J, Zhou T, Efros AA (2017) Image-to-image translation with conditional adversarial networks. In: 2017 IEEE conference on computer vision and pattern recognition (CVPR), pp 5967–5976https://doi.org/10.1109/CVPR.2017.632

Jiang Y et al (2022) Quo Vadis artificial intelligence? Discov Artif Intell

Kittler J (1983) On the accuracy of the Sobel edge detector. Image vis Comput 1(1):37–42

LeCun Y, Boser B, Denker JS et al (1989) Backpropagation applied to handwritten zip code recognition. Neural Comput 11(4):541–551

Liang Q, Miao Z, Geng J (2015) Weak-edge detection for motion capture using a two-layer active contour model. In: 2015 8th international congress on image and signal processing (CISP), pp 398–402. https://doi.org/10.1109/CISP.2015.7407912

Liu P, Wang G, Qi H, Zhang C, Zheng H, Yu Z (2019) Underwater image enhancement with a deep residual framework. IEEE Access 7:94614–94629. https://doi.org/10.1109/ACCESS.2019.2928976

Ma K, Xu Q, Wang B (2008) Roberts' adaptive edge detection method. J xi'an Jiaotong Univ 42(10):1240–1244

Naik A, Swarnakar A, Mittal K (2021) Shallow-UWnet: compressed model for underwater image enhancement. arXiv:2101.02073

Pizer SM et al (1987) Adaptive histogram equalization and its variations. Comput Vis Graph Image Process 39(3):355–368

Poma XS, Riba E, Sappa A (2020) Dense extreme inception network: toward a robust CNN model for edge detection. In: Proceedings of the IEEE/CVF winter conference on applications of computer vision

Shi Y, Yang W, Du H, Wang L, Wang T, Li S (2021) Overview of image captions based on deep learning. Acta Electron Sin 49(10):2048–2060

Tai S-C, Yang S-M (2008) A fast method for image noise estimation using Laplacian operator and adaptive edge detection. In: 2008 3rd international symposium on communications, control and signal processing, pp 1077–1081. https://doi.org/10.1109/ISCC.SP.2008.4537384

Torre V, Poggio T (2008) On edge detection. IEEE Trans Pattern Anal Mach Intell 8(2):147–163

Wang S et al (2017) Image crack detection with multiscale down-sampled normalized cut. Chin J Sci Instrum 38(11):2788–2796

Wang P, Chen P, Yuan Y et al (2018) Understanding convolution for semantic segmentation. In: Proceedings of IEEE winter conference on computer vision, pp 1451–1460

Wang X et al (2017) Edge preserving and multiscale contextual neural network for salient object detection. IEEE Trans Image Process 27(1):121–134

Wen H, Qi W, Shuang L (2016) Medical X-ray image enhancement based on wavelet domain homomorphic filtering and CLAHE. In: 2016 international conference on robots & intelligent system (ICRIS). IEEE

Xie S, Tu Z (2015) Holistically nested edge detection. In: Proceedings of the IEEE international conference on computer vision

Yang J et al (2016) Object contour detection with a fully convolutional encoder-decoder network. In: Proceedings of the IEEE conference on computer vision and pattern recognition

Zhu Q, Mai J, Shao L (2015) A fast single image haze removal algorithm using color attenuation prior. IEEE Trans Image Process 24(11):3522–3533

Multi-models Study on the Influence of Space–Time Factors on the Shared Bike Usage

Shangyang Liu, Lishan Yang, and Zhutao Zhang

Abstract Shared bicycles are an environmentally friendly and convenient means of transportation that can be found everywhere in life. However, problems such as over delivery, disorderly occupation and O&M mismatch may occur, which will seriously affect public order. In order to reduce the occurrence of such events, this study make models based on artificial intelligence technology, simulate the use of shared bicycles, and use random forest, xgboost studying, Long Short-Term Memory (LSTM) neutral network, artificial neural network and other algorithms due to their excellent ability for extracting features. The experimental results in this study demonstrated that all machine learning models achieved satisfactory results based on the collected dataset. In addition, Long Short-Term Memory neural network can achieve better performance compared to other models in terms of fitting results. Predicting the use of shared bicycles through artificial intelligence can provide new ideas for governments and shared bicycle companies to solve the problem of shared bicycles.

S. Liu, L. Yang, and Z. Zhang these authors contributed equally.

S. Liu (✉)
College of Electronic Engineering Technology, Electronic Information Science and Technology, Zhuhai College of Science and Technology, Zhuhai 519041, Guangdong, China
e-mail: 2691322395@stu.zcst.edu.cn

L. Yang
College of Software, Software Engineer, South China Normal University, Guangzhou 528225, Guangdong, China

Z. Zhang
College of Chemistry, Chemical Engineering and Materials Science, Soochow University, Suzhou 215031, Jiangsu, China

S. Yadav et al. (eds.), *Proceedings of 2nd International Conference on Artificial Intelligence, Robotics, and Communication*, Lecture Notes in Electrical Engineering 1063, https://doi.org/10.1007/978-981-99-4554-2_17

1 Introduction

Bicycle sharing is an emerging urban mode of transportation based on the theory of resource sharing, bicycle manufacturers to provide the general public with the use of Internet control of bicycles, the general public can use the convenience of the network in need of bicycles. In terms of the mode of sharing bicycles, there are two main issues. The first is the parking problem, due to the sharing of bicycles "no pile" operation, users often choose to park in any place, in real use, often cannot find a suitable parking place, so users often park at the subway station, bus platform, sidewalk, blind lane, and other public transportation areas, resulting in traffic congestion and other social problems. The second is that the overall saturation or surplus of bicycle delivery is excessive, but the regional bicycles companies that are in short supply and demand do not coordinate with the relevant government departments when designing the shared bicycle layout system, and because of the specific release, there will be delays and limitations, resulting in unbalanced bicycle layout and unreasonable delivery. In this case, predicting the use of shared bicycles based on advanced methods, e.g., machine learning algorithms is of great necessity. Predicting the results of the use of shared bicycles can provide a theoretical basis for the launch of shared bicycles by shared bicycle companies and can also provide new ideas for the government to solve the problem of shared bicycles.

In the early field of bicycle sharing business, the previous technology was mainly focused on traditional optimization algorithms, e.g., ant colony algorithm and simulated annealing algorithm, etc., which is different from the machine learning algorithm. Traditional optimization algorithms function well in areas closely related to spatial factors such as scheduling path optimization or managing shared bicycle parking area, but it also has disadvantages in processing more detailed information. Most of the theses using traditional optimization algorithms did not involve factors like weather, temperature, holiday and vacations, etc. For example, a public bicycle scheduling optimization based on simulated annealing algorithm considers only the data of existing parking place for shared bicycles without including the data of bike density in each parking place or other factors like weather (Yue 2021).

While papers using machine learning algorithm can involve multiple factors, for example, in a research of forecast of the demand for shared bicycles based on xgboost, time, season, holidays, working days, weather, actual temperature, body temperature, relative humidity, wind speed, and other environmental factors is involved in initial data analysis, and the result shows less error and higher accuracy compared with SVM model, though the time consumed is longer, xgboost still shows a promising outcome (Li 2021).

This research predicts shared bike delivery using multi-models. Not only limited in machine learning but also like deep learning, such as random forest, xgboost studying, Long Short-Term Memory (LSTM) neutral network, and artificial neural network. The procedure of deep learning and machine learning are essentially the same. From data set gathering, and then input features, training the model and finally get the prediction. Normal prediction of shared bike is based on traditional mathematical model. Replying to changes in time and environment is found hard. Traditional machine learning is not enough in this research. The efficiency of prediction is compared between different models. Initially, the data set is from 2012 Shared Bicycle Data Set in Washington Metropolitan Area. Under the imitation of the similar data set, the Shenzhen shared bicycle data set in 2021 was produced. The space features include such environmental factors as humanity, temperature, and wind speed. The time features are from year to hour. From previous research, multi-models and parameters adjustment is promising for the better bike-sharing algorithm which can be better than normal mathematical algorithm with low time consumption and high accuracy.

2 Method

2.1 Dataset

Data is collected from Hadi Fanaee-T. The number of rental bikes per hour and day between 2011 and 2012 was released by Porto on December 20, 2013 in the bicycle sharing system of the Portuguese capital, as well as the corresponding weather.

There are 16 labels in this dataset, and this study mainly uses 11 labels in Table 1.

The data related to weather is converted using one-hot coding to make it available for the model input. One-hot coding is used because most algorithms are calculated based on measures in vector space, so that the values of variables in non-partially ordered relationships are not partially ordered and are equidistant to the dots. Using one-hot coding, the value of a discrete feature is extended to Euclidean space, and a certain value of a discrete feature corresponds to a point in Euclidean space which

Table 1 Feature labels

Characteristic value	Empty or not	Data type
season	none	int64
mnth	none	int64
hr	none	int64
holiday	none	int64
weekday	none	int64
workingday	none	int64
weathersit	none	int64
temp	none	float64
atemp	none	float64
hum	none	float64
windspeed	none	float64

Table 2 The parameters of RF model

Param	Value
oob_score	True
max_depth	24
n_estimators	1500
n_jobs	−1
random_state	42

makes it easier for further process. One-hot encoding for discrete features is adopted to rationalize the distance between features.

In addition to this, we also normalize the data by feature data. The problem with numerical type variables is that each variable has a different range of changes and different units, and the solution we take is to standardize this variable and transform the data into a value that fluctuates in the [−1, 1] range.

2.2 Algorithms

In this study, linear regression, random forest, Xgboost, artificial neural network, and LSTM model are used in this study for training on the collected dataset. More details about these models except for the linear regression are described as follows.

2.2.1 Random Forest

Random Forest (RF) is a homogeneous ensemble method (Ve and Yongyun 2020). This approach is an integrated learning method for classification and regression proposed by Breiman. The idea is to combine huge decision trees with the same distribution. Each decision tree is created separately on the guidance data sample (Seo 2021). Observe the data again, the factor of weather is such a puzzled variable. For example, season is an expression of 1, 2, 3, and 4, which makes linear regression analysis inaccurate. According to the characteristics of such data, random forest can be considered as an effective method. Every tree in the forest depends on a random vector, and the vectors in the forest are independent and identically distributed. The final decision tree is generated based on the "vote" on the random vector potential tree, that is, the classification selection of the random forest gets the most votes. The parameters of RF model can be found in Table 2.

2.2.2 Xgboost

With optimized distributed gradient boosting library, Xgboost is designed to be efficient, flexible, and portable. It makes machine learning algorithm under gradient

Table 3 Parameters of xgboost

Parameters	Start value	Stop value	Best value	Optimum score
n_estimators	20	300	140	0.769
max_depth	3	10	9	0.81
min_child_weight	1	6	3	0.81
gamma	0	10	0	0.81
colsample_bytree	0	10	0.7	0.825
subsample	0	10	0.8	0.825
reg_alpha	3	5	3	0.832
reg_lambda	3	5	5	0.832
learning_rate	0.05	0.5	0.1	0.832

lifting framework to enhance the speed of the data training (Xgboost documents 2022). Besides, as a new integrated learning method, Xgboost supports parallel processing. As usual, it takes a lot of time to sort the values of feature decision tree training in common. Xgboost sorts the data in advance before training.

Generally, we need to choose a selective initial parameter. Learning rate is usually fluctuated in range (0.05, 0.3), we define learning rate (0.1) and then we turned the regularization parameters of model. Sophistication of models can be decreased by these parameters, in addition, we perceive that it promotes a lot in this model. In the process of determining a tree, we can select different parameters. Reducing the learning rate and determine the ideal parameters. The parameters of xgboost model can be found in Table 3.

2.2.3 Artificial Neural Network

Artificial neural network (ANN) is a computing model whose ideal and origin are inspired by biological neural network of nature human brain (Jain 1996; Krogh 2008). Artificial neural networks are very good at learning mapping relationships from input data and labels to complete prediction or solve classification problems, which has been demonstrated in many tasks (Yu 2022; Girshick 2015; Zhu 2023). Artificial neural network is also called universal fitting device because it can fit any function or mapping. Feedforward neural network has been used widely and frequently which makes it the most commonly used network. It generally includes three layers of artificial neural units, namely, input layer, hidden layer, and output layer. The hidden layer can contain multiple different layers, which constitutes the so-called deep neural network.

(1) It can be seen from the data pre-processing above that under the one heat coding, the data is divided into 56 features, so the input layer of the neural network is 56 neurons, and the hidden layer is artificial prediction, which is adjusted to 10 neurons this time.

(2) The configuration of batch size. In the training cycle, we will also encounter a problem. In the previous example, in each different training cycle, we input all the data into the neural network. There is no problem when the amount of data is not large. However, the current data volume is 16875. In such a large data volume, if all the data are processed in each training cycle, the operation speed will be too slow and the iteration may not converge. The solution is usually to adopt the batch processing mode, that is, divide all data records into a small batch size dataset, and then input a batch of data to the neural network in each training cycle. The batch size depends on the complexity of the problem and the size of the data volume. In this example, the batch size is set to 128.

2.2.4 LSTM

Long short-term memory model is classified as a special RNN model, which is designed and coded to solve the problem of gradient dispersion of RNN model; in traditional RNN model process, by contrast, if the time is long, the number of the residual error returned will decline exponentially, as a result, the weight of network is updated slowly, which cannot reflect the long-term memory effect of RNN. Therefore, a storage unit is required to store memory, so LSTM model is proposed; LSTM model is a time recurrent neural network, which befits for dealing with and predicting considerable events with relatively long intervals and delays in time series.

(1) Since the LSTM is a type of Neural Network, the number of neutrons is set to 56 in this study.
(2) The first layer of Neural Network used traditional activation "relu," input_size is the size of input data. We choose "return_sequences" to be true. For the second layer of the model, we halve the input size and still select "relu" activation.

3 Result

There are five models used in this prediction, including linear regression model, random forest, LSTM, xgboost model, and artificial neural network model.

3.1 Linear Regression Model

For the linear regression model, the result has a root mean square error of 20,000 with unprocessed data and 12,000 with processed data, which is awful. So linear regression model is eliminated in the early part of this research.

Table 4 Result of the random forest

Training set quantity	Test set quantity	MSE	MAE	RMSE	Explained variance	Out of bag (OOB)	r2_ score	Average process time
16,659	720	0.2	0.311	0.447	0.775	0.917	0.756	10.9 s

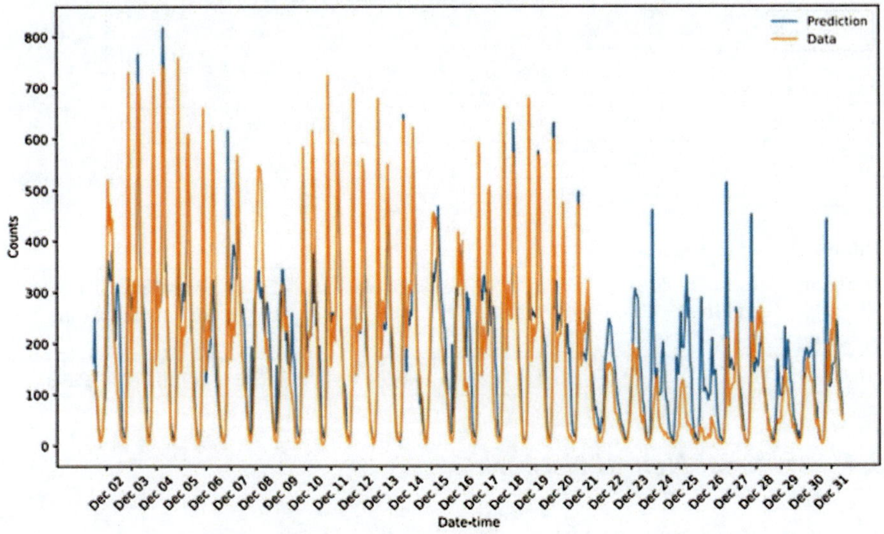

Fig. 1 Overview of forecast results for random forest

3.2 Random Forest

Random forest is introduced due to season, weather is dummy variable factor. For example, season is expressed intermittently in the form of season 1, season 2, season 3, season 4, which makes the linear regression analysis appears to be inaccurate. The result is shown in Table 4 and Fig. 1.

3.3 Xgboost Model

Xgboost can use parameter adjustment to improve the gradient. However, after multiple parameter adjustments, the model fails to meet expectations. However, understanding the meaning behind each parameter can help train a better model. The related result is shown in Tables 5 and 6.

Table 5 Value and score parameter

Parameter	Initial value	Final value	Best score value	Best score
n_estimators	20	300	140	0.769
max_depth	3	10	9	0.81
min_child_weight	1	6	3	0.81
gamma	0	10	0	0.81
colsample_bytree	0	10	0.7	0.825
subsample	0	10	0.8	0.825
reg_alpha	3	5	3	0.832
reg_lambda	3	5	5	0.832
learning_rate	0.05	0.5	0.1	0.832

Table 6 Result of xgboost model

Training set quantity	Test set quantity	MSE	MAE	RMSE	Explained variance	r2_ score	Parameter regulation time	Training time
16,659	720	1472.4	16.9	38	0.79	0.771	13 min	2.3 s

3.4 Artificial Neural Network

The related results of the ANN model can be found in Fig. 2, Table 7, and Fig. 3.

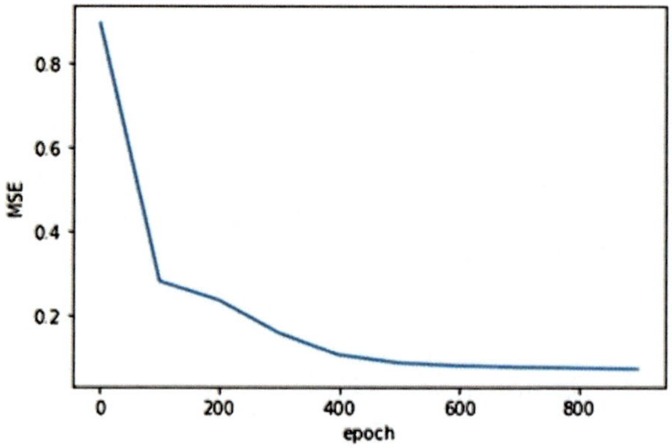

Fig. 2 The relationship between average error and training time

Table 7 Result of artificial neural network

Training set quantity	Test set quantity	MSE	MAE	RMSE	Explained variance	r2_score	Training time
16,659	720	0.2	0.28	0.42	0.84	0.793	74 s

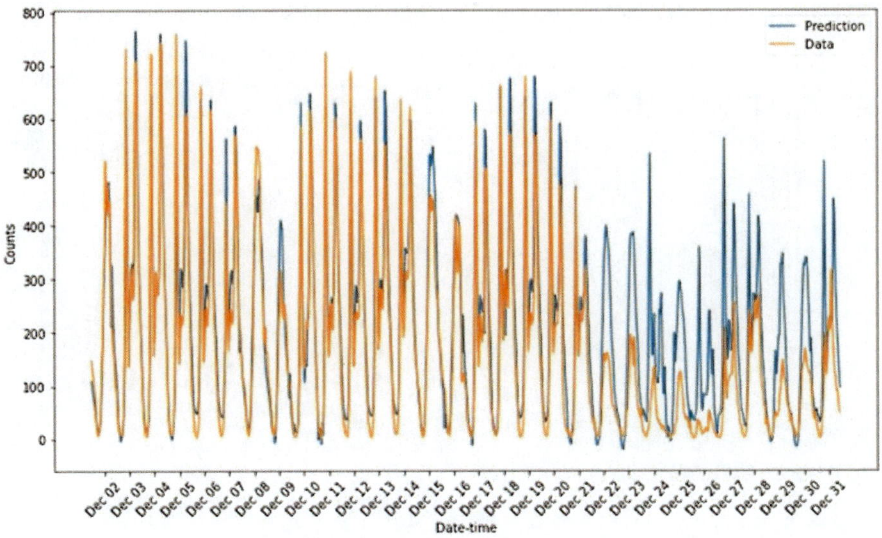

Fig. 3 Overview of forecast results for artificial neural network

Table 8 Comparison of random, xgboost model, and artificial neural network

Model name	Training set quantity	Test set quantity	MSE	MAE	RMSE	Explained Variance	r2_ score	Average process time
Random forest	16,659	720	0.2	0.311	0.447	0.775	0.756	10.9 s
Xgboost model	16,659	720	1472.4	16.9	38	0.79	0.771	13 min/ 2.3 s
Artificial neural network	16,659	720	0.2	0.28	0.42	0.84	0.793	74 s

3.5 *Comparison of Random, Xgboost Model, and Artificial Neural Network*

The comparison based on the several models in terms of the performance can be found in Table 8.

3.6 *LSTM Long and Short-Term Memory Model*

The results of the LSTM model can be found in Fig. 4.

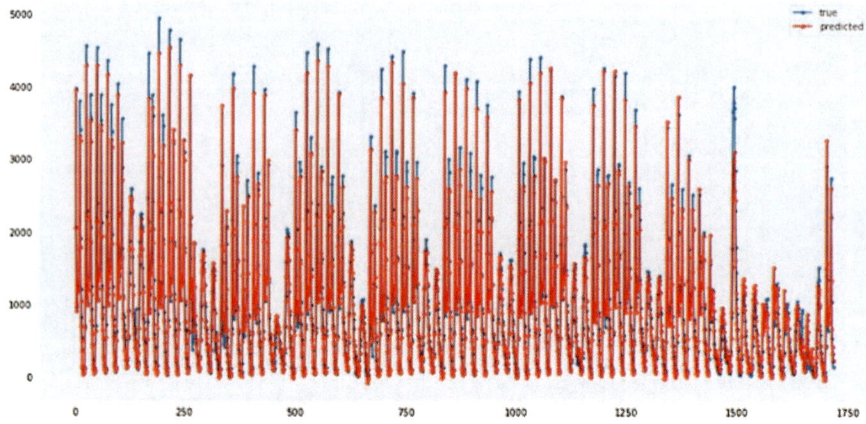

Fig. 4 Overview of forecast results for LSTM

4 Discussion

Among five models, linear regression model was eliminated first due to its high error, but its elimination reflects that the relationship between the data set characteristics and the result set is not a simple linear relationship. The particularity of data form in time series prediction leads to various pitfalls in the process of building models. If it is a single point prediction, LSTM can achieve very good results. Another consideration is that the data noise is not large, and the noisier the data set is, the better the effect will be obtained by using a simpler model, while the effect will be poor by using a complex model due to over fitting.

5 Conclusion

In this study, the performance of several machine learning models on influence of space–Time factors on the shared bike usage is investigated. Some typical models, e.g., random forest, xgboost studying, Long Short-Term Memory (LSTM) neutral network, artificial neural network were considered for predicting the collected dataset. The experimental results indicated that the LSTM model can achieve more satisfactory performance than other models. In the future, some more advanced models may be considered to further improve the accuracy of the model.

References

Girshick R (2015) Fast r-cnn. In: Proceedings of the IEEE international conference on computer vision. pp 1440–1448

Jain AK (1996) Artificial neural networks: a tutorial computer 29(3):31–44

Krogh A (2008) What are artificial neural networks? Nat Biotechnol 26(2):195–197

Li F, et al (2021) Prediction of the demand for shared bicycles based on XGBoost algorithm. J Wuhan Univ Technol (transport science and engineering edition) 45(5):880–884

Seo Y (2021) Predicting demand for a bike-sharing system with station activity based on random forest. In: Proceedings of the institution of civil engineers-municipal engineer, vol 174. no 2. Thomas Telford Ltd

Ve S and Yongyun C (2020) Season wise bike sharing demand analysis using random forest algorithm. Comput Intell

Xgboost documents (2022) https://xgboost.readthedocs.io/en/stable/

Yu Q (2022) Pose-guided matching based on deep learning for assessing quality of action on rehabilitation training. Biomed Signal Process Control 72:103323

Yue X (2021) Public bicycle scheduling optimization problem based on simulated annealing algorithm Science and Technology. Innovation 36:19–21

Zhu M (2023) Investigation of nearby monitoring station for hourly PM2. 5 forecasting using parallel multi-input 1D-CNN-biLSTM. Expert Syst Appl 211:118707

Application of Collaborative Robot in Cigarette Production Line for Automatic Distribution of Packaging Materials

Du Jing, Dao Ronggui, Wu Peng, Zhang Yongshou, Wu Bogang, and Wu Guiwen

Abstract At present, AGV is used to distribute packaging materials in cigarette production line. This method occupies a large space around the production unit, and a large number of personnel and forklifts participate in material distribution, resulting in chaotic production area. Therefore, the transportation collaborative robots are considered. The transportation collaborative robot is mainly composed of a vehicle, a driving part and an executive device. It combines the following design elements to form an organic integrated logistics system, such as the transport unit, the production consumption and supply matching in the distribution process, the intelligent material station, and the robot collaborative distribution operation process. The system is applied to the automatic distribution of packaging materials in the cigarette production line. After the collaborative robot carries the material pallet to the production unit, it can also automatically distribute the coil type materials. The system realizes automatic feeding for the production unit, reduces the amount of manual labor, and improves the level of site management.

1 Introduction

In China, the tobacco industry is the pioneer in the application of industrial automation equipment. In the past 30 years, the research and application of automation technology in cigarette factories have never stopped. So far, its technical equipment has reached the international advanced level. AGV (Automatic Guided Vehicle) [1, 2], robot [3, 4] and other automatic equipment are used in cigarette production line, respectively for material handling or selection [5, 6]. But the combination of the characteristics of the mobile manipulator can make the production line process more streamlined and efficient, obviously more flexible. There is no case of intelligent distribution of cigarette materials with such equipment. With the rapid development

D. Jing · D. Ronggui (✉) · W. Peng · Z. Yongshou · W. Bogang · W. Guiwen
Hongta Tobacco (Group) Co., Ltd., Yuxi 653100, Yunnan, China
e-mail: 664188929@qq.com

S. Yadav et al. (eds.), *Proceedings of 2nd International Conference on Artificial Intelligence, Robotics, and Communication*, Lecture Notes in Electrical Engineering 1063, https://doi.org/10.1007/978-981-99-4554-2_18

and maturity of new technology, the possibility of changing the distribution mode at the end of the logistics supply chain is increasing. The realization of intelligent distribution, unmanned distribution, self-learning distribution model will no longer be just imagination, but will become a reality. These new technologies include intelligent collaborative robotics, visual recognition, big data and cloud platforms. Today, we will study the intelligent distribution mode of cigarette production logistics materials.

2 Existing Problem

Let's look at the current cigarette production line distribution mode of packaging materials. When the materials in the line side warehouse are about to be exhausted, the production personnel call the required materials in the operating terminal computer. The automated logistics system takes out the production materials from the elevated warehouse and sends them to the line side warehouse. The transportation process adopts automatic logistics equipment such as tunnel stacker, conveyor and AGV.

There are more problems and shortcomings. First, the pallet type distribution should be set up next to the production unit according to the material specifications. Lead to a variety of materials, occupy a large space around the production unit, the unit is crowded and messy. Secondly, a large number of personnel and forklifts are involved in the distribution of materials. The production area is cluttered, there are big security risks, and the distribution efficiency is low. This situation is not conducive to the overall operation of intelligent logistics. As shown in Fig. 1.

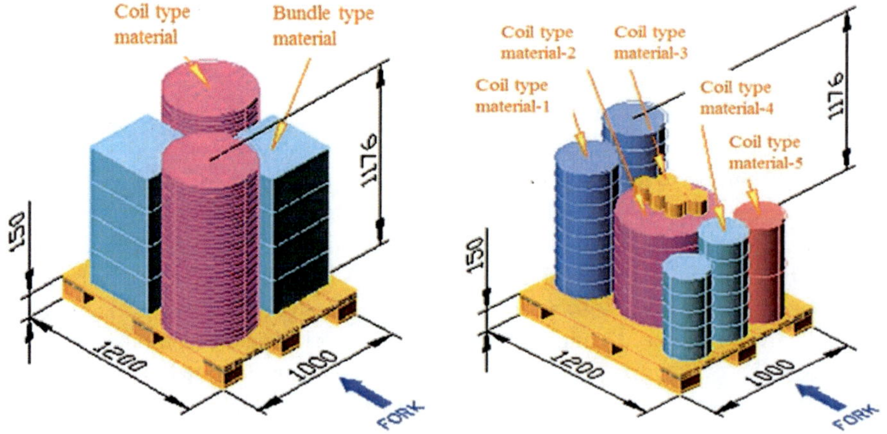

Fig. 1 Current production unit line-side warehouse pallet mode

3 Analyzing Problem

3.1 Improvement Ideas

Explore the robot intelligent collaborative mode of cigarette production line automatic distribution of packaging materials, improve the automation and intelligent level of material distribution, the goal is to build an unmanned factory. The feasibility method is as follows: in the material supply chain, the pallet is taken as the minimum transport unit, and in the use terminal such as the cigarette production line, intelligent distribution technology is adopted to decompose the minimum transport unit into the minimum use unit. Through data analysis and reasonable matching of production consumption, the production line can be delivered on demand, timely and accurately.

3.2 Data Analysis

Packaging materials can be divided into three categories according to their shape:

3.1.1. *Coil type material.* Coil type material has 19 specifications, outer diameter size: ϕ 135 mm–ϕ 595 mm, inner diameter size: ϕ 30 mm–ϕ 150 mm, thickness: 18.5–345 mm, weight: 1.15–25 kg.

3.1.2. *Barrel type material.* Barrel type material has 2 specifications, outer diameter size: ϕ 317 mm, height: 380 mm, weight: 20–25 kg.

3.1.3. *Bundle type material.* Bundle type material 3 specifications, length: 48–250 mm, width: 22–100 mm, height: 35–410 mm, weight: 0.092–6.625 kg.

From the classification and specification data statistics, the packaging materials are mainly coil type, accounting for 67.86%, close to 70%, and the consumption is relatively large. The distribution of coil type materials should be regarded as the key point to solve the problem, which is the breakthrough direction to realize the intelligent collaborative distribution mode.

4 Improved Method

Design intelligent collaborative distribution mode of packaging materials in cigarette production line.

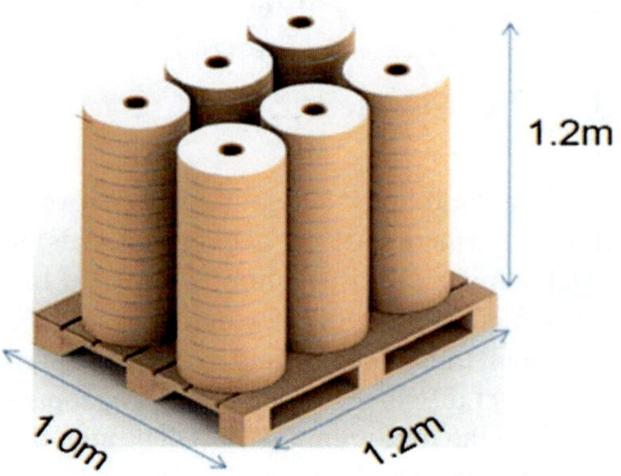

Fig. 2 Schematic diagram of material pallet full load dimensions

4.1 *Transport Unit Design*

Pallet standardization in the transport unit of the material supply chain needs to consider many factors, including size, weight, outer packaging form, transport vehicle loading mode, automatic loading and unloading, etc. This not only facilitates transportation and storage, but also does not waste equipment capacity. At the same time, it can improve the efficiency of circulation and handling links and reduce logistics costs. According to the present elevated warehouse, the full load weight of the pallet should be less than 1000 kg, and the length, width and height size should not exceed L * W * H = 1200 * 1000 * 1200 (mm). As shown in Fig. 2.

4.2 *Matching Design of Production Consumption and Supply Quantity in Distribution Process*

Material distribution supply is related to many factors: production unit model, production capacity, material consumption per unit time, material type, material unit consumption time, moving robot walking speed, robot loading and unloading beat, robot electric energy supplement, etc. Based on the above factors, the unit time supply of each production model can be matched for reasonable distribution, which not only meets the production demand, but also takes into account the cooperative robot handling efficiency and reduces the equipment investment.

4.3 *Collaborative Robot Design*

As a transportation collaborative equipment, the mobile manipulator mainly distributes coil type materials. Fixture to achieve miniaturization and lightweight.

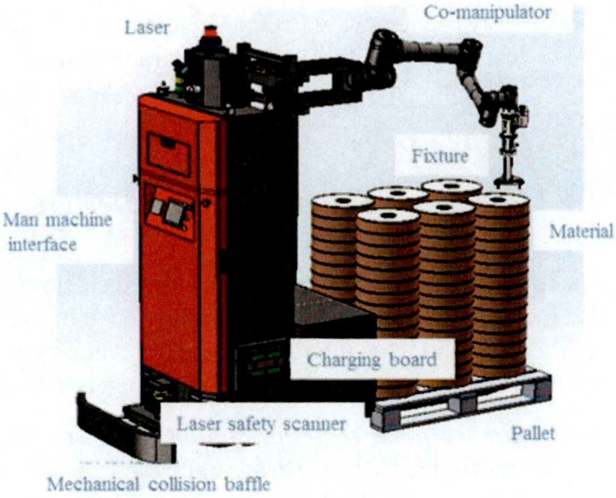

Fig. 3 Schematic diagram of transportation collaborative robot used in cigarette production line

With the help of visual identification system, it can quickly locate and move objects, correct offsets, and pick up materials adaptively. The transportation collaborative robot is mainly composed of a vehicle, a driving part and an executive device. As shown in Fig. 3.

4.3.1 *The vehicle.* It's forklift style suitable for the transport of material standard pallet, compatible with the existing elevated stereo library standard pallet storage unit.

4.3.2. *The driving part.* It's steering wheel design for semi-independent suspension mode. This can increase the amount of contact between the steering wheel and the ground, to solve the problem of driving wheel skid caused by uneven ground. The steering gear is driven by AC motor. Compared with DC motor, it has the advantages of higher control precision, faster response speed, low noise and no need to replace the carbon brush.

4.3.3. *Execution device.* It is composed of a cargo fork and a cooperative manipulator. The cooperative manipulator is equipped with a visual identification camera, which can accurately identify the center of the coil type materials. It can grab materials with precision. The cargo fork has the lifting function. This can butt with the conveyor, and fork the pallet.

4.3.4. *Main parameters.* Control system: NDC8; Guidance mode: laser guidance; AGV: customized forklift; Load shifting mode: fork and cooperative manipulator; Control mode: manual or automatic; Body size : L * W * H = 2500 mm * 800 mm * 2500 mm; Obstacle crossing, slope climbing and seam crossing capacity: ≤10 mm groove, ≤2°; Load weight: ≤1000 Kg; Lifting mode: hydraulic; Maximum lifting height: ≥1200 mm; Walking speed: forward Vmax ≥ 60 m/min, backward Vmax ≥30 m/min, acceleration (Max) 0.5 m/s^2; Navigation accuracy: ±10 mm; Stop accuracy: ±10 mm; Manipulator grasp accuracy: ±2 mm; Vehicle body clearance height: 30 mm; Communication mode: wireless LAN; Grab the maximum weight ≤11 KG; Fixture: custom; Coverage: whole tray (L * W = 1.2 m * 1.0 m).

4.4 Intelligent Material Station Design

The transportation collaborative robot needs to unload coil type materials to a specified location to accommodate the production unit's automatic equipment or personnel access. Two kinds of intelligent material stations are designed, one is rotary plate material station and the other is cabinet material station.

4.4.1. *Rotary plate material station.* It is set next to the production unit and controlled by PLC (Programmable Logic Controller). Photoelectric lighting is installed in the material station to detect material consumption. The material station carries two kinds of materials, one main and one standby storage, and the consumption ratio is determined according to the demand of the unit. When the material of side A is exhausted, the photodetector will trigger the material signal, the turntable will rotate to side B, and the unit operator will continue to work. The PLC sends the address block to the WCS (Warehouse Control System) system to produce the material task, and the transportation collaborative robot carries the material to the unit for the material replenishment, and the material is replenished quantitatively each time. As shown in Fig. 4a.

4.4.2. *Cabinet material station.* It is also located next to the production unit. Heavy materials are stored in the lower layer and light materials are stored in the upper layer. Each bin of the cabinet is equipped with a set of detectors and a

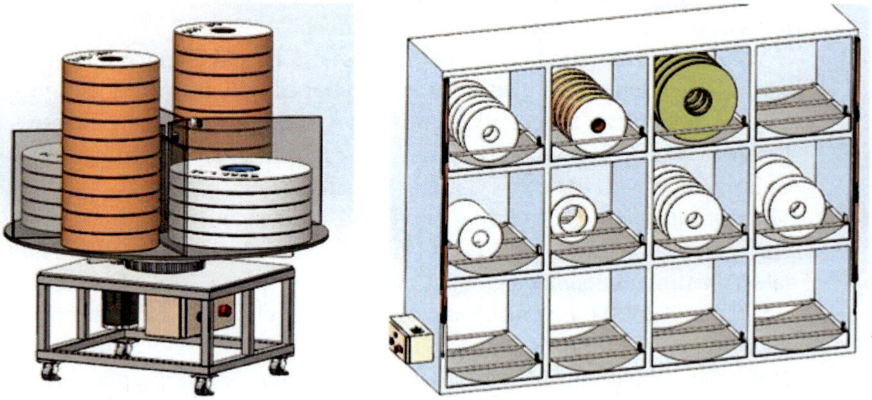

(a) Rotary plate material station (b) Cabinet material station

Fig. 4 Intelligent material station

(a) Mark for placing materials in the center pallet (b) Arrange materials with guide posts

Fig. 5 Transport stability design

safety grating. When manual feeding, the transportation collaborative robot stops feeding at the express cabinet. when the workers take the materials away from the safety grating, the transportation collaborative robot continues the feeding operation. The feeding of each bin is quantitative. As shown in Fig. 4b.

4.4.3. *Transport stability design.* The collaborative robot transports pallet of materials from the storage area to the production unit, and the transport process needs to be stable. Draw a center cross line on the pallet to balance the gravity of the material placement. As shown in Fig. 5a. When placing the coil type material, use the circular guide column to make the material neat and not fall off. As shown in Fig. 5b.

4.4.4. *Handling stability design.* The collaborative robot moves the coil type materials from the vehicle to the material station, which needs to be prevented from falling. The pneumatic gripper with three claws is designed. Each claw is sawtooth cross shape to ensure that the sawtooth embedded in the inner diameter of the material does not fall off. As shown in Fig. 6a. The effect of collaborative robot grasping material is shown in Fig. 6b.

4.5 Robot Collaborative Distribution Operation Process Design

4.5.1. *Outbound process.* After the logistics system produces the outbound task, the material pallet will be transported to the unpacking platform, and the external packaging of the pallet will be removed manually or automatically and then sent to the outbound connection platform. The collaborative robot will move it to the production unit, and the vehicle-mounted manipulator will grab the corresponding amount of materials to the specified material station.

(a) Diagram of claw tool (b) Use effect of claw tool

Fig. 6 Handling stability design

4.5.2. *Pre-production material distribution.* According to the daily production plan and material regulation, the unit is divided into groups. There is no material on the platform of the unit when the shift starts. MES (Manufacturing Execution System), WMS (Warehouse Management System) and other systems automatically generate outbound tasks and assign them to the WCS system. The WCS decomposes the tasks and assigns them to each single system for execution. The collaborative robot will take down the material pallet and transport it to the corresponding unit. A certain amount of materials will be placed on the material station through the vehicle-mounted manipulator. The first distribution of materials will meet the consumption of the unit for at least 2 hours. After the delivery of one station, the collaborative robot will automatically run to the next unit to distribute materials. When the material pallet empty, the collaborative robot will automatically send it to the recycling platform.

4.5.3. *Replenishment distribution in production.* The detection switch of the material station will automatically trigger the replenishment task after the machine consumes the amount of material for 1 hour. MES, WMS and other systems will automatically generate the task of discharging replenishment. The collaborative robot feeds the unit. When other units need feeding, the collaborative robot directly runs to the target unit for feeding. Without feeding, the system commands the intelligent distribution robot to prepare for the next distribution task through calculation and data prediction.

4.5.4. *Surplus material storage and distribution.* When there is surplus material on the pallet, the collaborative robot will apply for a new unloading address and run to the new unit for unloading. If the WCS system does not deliver a new address, it unloads the surplus pallet to the specified cache location. When there is a new distribution demand, the priority is to pick up the materials from the surplus cache location. After the materials on the pallet are used, the collaborative robot will retrieve the empty pallet directly.

5 Conclusion

The application of transportation collaborative robot in cigarette production line for the automatic distribution of packaging materials has changed the mode of AGV in the previous production line which only carries pallets, not materials. The application of the material station can automatically replenish the material for the production unit, reduce the labor amount of manual inspection of the line storage and manual material, and reduce the production workshop where the material is placed, and improve the level of site management.

The application of transportation collaborative robots can greatly reduce the repeated handling operations in the supply chain, avoid material damage, ensure material quality and reduce logistics costs. The transportation collaborative robot distribution mode solves the differences and contradictions between warehousing and production in the material supply business. The handling stability of cooperative robots can reduce the use of external packaging of materials, reduce procurement costs, and help environmental protection, promote the management level of low-carbon green development, and enhance the corporate image.

Acknowledgements The research is supported by the following funds: 2021 Hongta Tobacco (Group) Co., Ltd., China Tobacco Yunnan Industrial Co. Ltd. science and technology project *Research and application of intelligent collaborative distribution mode based on cigarette materials robot*. Project number: 2021JC10.

References

Duan Q, Li H, Zhang C et al (2018) Design of intelligent cigarette material dispensing and pallet loading system. Tob Sci Technol 51(8):100–104, 108

Jiang H (2017) Improvement of AGV system in cigarette ingredient warehouse. Logist Technol 36(9):150–153

Lei C, Jie L (2018) Research on intelligent pallet loading of cigarette material in transition to flexible production. Logist Technol 37(10):105–109

Li C, Wu G, Liu F et al (2010) Optimization of AGV system in tobacco industry. Tob Sci Technol 11:18–21

Linchao Y, Xinfeng Z (2021) Design and control of clamping device of cigarette product palletizing robot. Mach Des Manuf Eng 50(11):61–64

Yang J, Xu Z, Zhao Y et al (2018) Application of robot for automatic case loading system in cigarette case filling and sealing machine. Tob Sci Technol (2):20–23

Comparison of Data Processing Performance of Hadoop and Spark Based on Huawei Cloud and Serverless

Shize Pang, Runqi Su, and Ruochen Tan

Abstract The optimal choice between serverless computing and traditional server computing, the current frontier in the machine learning and cloud server industries, has always been controversial. In contrast, serverless computing has better resource utilization but is restricted to a certain range of applications. Therefore, the research on serverless computing is not only the future direction of development, but also an essential practice direction. This paper deployed and built an experimental solution based on Huawei Cloud serverless platform, compared and analyzed various data processing algorithms under MapReduce in Huawei Cloud serverless platform, summarized their advantages and disadvantages, and found the areas where each algorithm is promising.

1 Introduction

Machine learning is a branch of artificial intelligence that studies and analyzes how to make computers capable of human learning behavior through knowledge from different interdisciplinary disciplines. As a branch of artificial intelligence, machine learning is one of the ways people can reach artificial intelligence. Artificial intelligence (AI) can dominate the process of machine learning, but the result of machine learning does not necessarily lead to artificial intelligence. On the other hand, machine

Shize Pang, Runqi Su and Ruochen Tan these authors contributed equally.

S. Pang
School of Computer Science (National Pilot Software Engineering School), Beijing University of Post and Telecommunications, Beijing 100876, China

R. Su (✉)
School of Computer Science and Information Engineering, Xiamen Institute of Technology, Xiamen 361021, Fujian, China
e-mail: p2311301@mpu.edu.mo

R. Tan
School of Communication, University of Miami, Miami, FL 33124, USA

S. Yadav et al. (eds.), *Proceedings of 2nd International Conference on Artificial Intelligence, Robotics, and Communication*, Lecture Notes in Electrical Engineering 1063, https://doi.org/10.1007/978-981-99-4554-2_19

learning algorithms are the core technology that provides the machine with the ability to think logically on its own and improve itself through experience. Researchers typically categorize machine learning algorithms into supervised and unsupervised learning, and of course, the emerging concept of enhancing potential learning. Supervised learning is one sub-category of machine learning and artificial intelligence. It is defined as the use of labeled datasets to form data classification algorithms or to precisely predict results. As input data is entered into the model, the model adjusts its weights until the model is adapted, and this happens as part of the cross-validation process. Supervised learning helps organizations solve a variety of real-life issues on a large scale, like sorting spam from your inbox into a separate folder. Unsupervised learning is an automatic learning type. Machine learning refers to an AI (Artificial Intelligence) subset where AI can learn to become more intelligent over time. With unsupervised learning, professionals first deliver existing datasets to AI. Artificial intelligence then analyzes the trends of these data sets. AI creates rules that help them understand and categorize data according to trends they detect. As more data becomes available to AI, their analysis and rules become more rigorous and precise. When an AI uses unsupervised learning, a professional provides the initial data and then the AI is responsible for determining the outcome.

This paper specifically studies the data processing part under machine learning (Grossi et al. 2107; Kuśnierz et al. 2022). As the root content of machine learning, data is not only the guarantee that computers can learn continuously and have human-like thinking patterns, but also the main way and means to optimize and continuously improve machine learning algorithms nowadays. This paper selects two popular and controversial frameworks, Hadoop and Spark, and executes the Wordcount algorithm of MapReduce model under them, and records and organizes the specific parameters under different text volumes in real time, so as to find out and analyze the selection and use of Hadoop and Spark frameworks in specific situations.

2 Methods

2.1 Hadoop

Hadoop is a user-friendly open-source distributed system infrastructure framework developed and built by the Apache Foundation. Its main work and use lie in the storage of large amounts of data and the processing of large data sets in large quantities through the parallel operation of many server clusters. As one of the mainstream distributed architectures for big data processing, its development and integration can be traced back to Apache's Nutch project in 2002. Based on the vision of MapReduce computing architecture in Jeff Dean and Sanjay Ghemawat's paper "MapReduce: Simplified Data Processing on Large Clusters," Doung Cutting and his development team first combined the MapReduce algorithm with Nutch Distributed File System (NDFS), which is the prototype of the subsequent Hadoop framework. In the whole

Hadoop framework, there are four important components: (1) Hadoop Common, a module that encapsulates and integrates common underlying tools, is the underlying logic of Hadoop, which mainly contains parameter configuration tools, remote procedure calls, program serialization tools and Hadoop's own abstract file system management tools and is the cornerstone of other Hadoop modules. (2) Hadoop Distributed File System (HDFS). A file distributed management system for storing file data in a multi-server cluster (Shvachko et al. 2010). It locates the location of specific files mainly through a directory tree. Because it is a distributed management system, it needs to be federated between servers to achieve functionality and is suitable for single write multiple output scenarios and does not support file modification (Thain et al. 2005). (3) Hadoop Yet Another Resource Negotiator (YARN). A multi-server resource provisioning tool, often playing the role of coordination and allocation between tasks and characters. The core idea is to make the two main tasks of JobTracker independent (resource management and job scheduling and monitoring), thus relieving the pressure of JobTracker alone and making the whole macro resource provisioning more distributed and unified (Vavilapalli et al. 2013). (4) Hadoop MapReduce, a programming model derived from the partitioning method, whose main functions are Hadoop MapReduce is a programming model derived from the partitioning method, whose main function is to slice and pack each individual data file into individual machines for data processing, and then combine and pack the results of each machine to get the desired result. Compared to other distributed architectures for Big Data processing, Hadoop is enduring because of its powerful data storage and processing capabilities (which can be enhanced in a simple way) and its excellent scalability (the open-source architecture allows each company to customize and develop based on its standard solution).

2.2 MapReduce

MapReduce is a program model based on the Hadoop framework developed by users through distributed computing. It is also one of the core components of Hadoop, and its origins can be traced back to Jeff Dean and Sanjay Ghemawat's 2003 paper "MapReduce: Simplified Data Processing on Large Clusters," which simplified the complexities of Hadoop and abstracted the two general and understandable models of Map and Reduce (Dean and Ghemawat 2008). The core function of MapReduce is to organize the user's own custom code and environment components into a complete program and run it concurrently on a Hadoop cluster. A complete and independent MapReduce program runs with three instances, each of which exists independently as a process (Giménez-Alventosa et al. 2019). (1) MRAppMaster, which is responsible for handling the entire program process and resource provisioning, as well as coordinating the independent state control of each program. (2) MapTask, which is mainly responsible for the source data entering the map phase of the MapReduce program. (3) The core of MapReduce lies in the Map phase and the otherworldly Reduce phase. In layman's terms, the Map process is a process of processing and chunking the source data, while at the same time filtering out unwanted or incorrect data content to achieve the effect of dynamic data filtering. In contrast to the Map phase, the Reduce phase is more like a process of batch processing and simplifying the sorted and filtered data. In addition, MapReduce has the following advantages: (1) Versatility. Users can write code programs and deploy them to the MapReduce model in a variety of programming languages, such as Java, C++, and Python. (2) Excellent resource scalability. Whenever our computational resources are not available, we can expand and extend its computational power by simply adding more machines to process the data, which is often simple and practical. (3) Highly sophisticated fault tolerance mechanisms. Since MapReduce is designed to be deployed on a cluster of inexpensive servers, if one of the servers fails to function properly, MapReduce's provisioning mechanism will transfer the computational tasks above to another node, allowing each task to function properly. In addition, this process does not require manual human involvement, but is completely automated and executed internally by Hadoop. (4) Suitable for offline processing of large-scale data. Due to MapReduce's unique parallelism mechanism, it can be deployed and implemented on a server cluster of thousands of servers to process massive amounts of data at the petabyte level or more, providing good data analysis capabilities. The experimental results in this article are all based on MapReduce under Hadoop 2.X version, since the version of Hadoop 2.X is still heavily used nowadays.

2.3 Spark

Apache Spark is a distributed open-source processing system which was designed for big data workloads (Zaharia et al. 2016). It uses cached and query execution, which is optimized in the memory, so that any size of data could be analyzed quickly. It provides plenty of development APIs in different languages such as Java, Scala, and Python. The original purpose of create Spark is there are so many limitations in MapReduce. It can perform memory processing, reduce the steps in the job, and reuse data across multiple parallel operations. Spark takes only one step to read data into memory, perform operations and write back, which greatly improves the efficiency of execution (Castro et al. 2019).

Spark has five features: (1) Fast—use state-of-the-art DAG schedulers, query optimizers, and physical execution engines to provide high performance for batch processing and streaming data. (2) Easy to use—it helps to write applications using Java, Scala, Python, R, and SQL. It also provides more than 80 advanced operators. (3) Generality—It provides a series of libraries, including SQL and DataFrames, MLlib, GraphX, and Spark Streaming for machine learning. (4) Lightweight—it is a lightweight unified analysis engine for large-scale data processing. (5) Ubiquitous—it can easily run-on Hadoop, Apache Mesos, Kubernetes, standalone, or in the cloud.

2.4 Secure File Transfer Protocol

Secure File Transfer Protocol (SFTP) is a secure file transfer protocol, a secure way to transfer files over the network (Durve and Mhatre 2015). It ensures the safe transmission of data using private and secure data streams. SFTP requires client users to be authenticated by the server, and data transmission must be carried out through a secure channel (SSH), that is, plaintext passwords or file data are not transmitted. SFTP does not use separate command channels and data channels, but data and commands are transmitted in a special format of packets through a single connection. It allows various operations on remote files, which is a bit like a remote file system protocol. SFTP allows recovery from operations such as pause transmission, directory list, and remote file deletion.

3 Experimental Results and Analysis

The experiment is based on Huawei Cloud Elastic Compute Service (ECS) Server, including four pay-as-you-go instance hosts on Huawei Cloud, each with specifications of 2 vCPUs, a memory of 4GiB, a CPU of Huawei Kunpeng 920, a storage capacity of 40 GB, a maximum bandwidth of 3 Gbit/s, and intranet sending and receiving packets of 300,000 PPS. The system selected is CentOS 7.6 64 bit with ARM under Linux, and the four instances are named TX-0001-0001, TX-0001-0002, TX-0001-0003, and TX-0001-0004. Besides, this paper uses Huawei Cloud's Object Storage Service (OBS) and creates storage buckets. The storage bucket is a container for storing objects in OBS and is used to store the author's data.

The installation and testing of Hadoop and Spark clusters in the preparation phase of the experiment. Firstly, two ports, 8020 and 50,010, are added to Huawei Cloud Security Group, with priority 1. Port 8020 is opened to access Hadoop, and port 50,010 is opened to write data. Next, this paper sets the IP mapping of the four instances in the /etc./hosts file. Then the firewall was turned off and disabled from booting. This paper then generates RSA-type SSH keys on all four instances, aggregates the id_rsa.pub files of the four instances, and copies them to the authorized_keys of each of the four nodes. The first connection to each node was verified and was successful. This paper chose Hadoop version 2.7.7 and JDK version OpenJDK8U-jdk_aarch64_linux_openj9_8u292b10_openj9-0.26.0. This paper then installed Hadoop and JDK on each node and configured the host and environment variables.

This paper choses Spark version spark-2.1.1-bin-hadoop2.7. After installing and unpacking the Spark installation package, edit the configuration file to add variables. Then restart the Hadoop cluster. Finally, verify the success of Spark installation with the command.

This paper installed Zookeeper, Hbase, Hive, and MySQL to run Spark. Zookeeper version 3.4.6, Hbase version 2.0.2, and MySQL version 5.7.30 were selected. After configuring the files, start Zookeeper and Hbase, download the MySQL installation package on the master node, install the dependencies needed for MySQL, and finally unzip the MySQL installation package. Then, enter the aarch64 directory, install the rpm package, and start the MySQL service. The default encoding configuration is utf-8. The Hive cluster version chosen by the author is 2.1.1. Create the Hadoop user in MySQL after downloading and unpacking (password: Hadoop). After that, create the database connection. After configuring the Hive files, copy the MySQL connection driver to the lib directory in the hive root directory. Then this paper configures the system zkpk user environment variables, starts and verifies the Hive installation, modifies the Hadoop cluster configuration, and finally turns on Hive remote mode.

After completing the basic construction and preparation of the Hadoop and Spark platforms, this paper examined and filtered the data content that needed input. Since the next research is based on the MapReduce model, considering that the key process Map and Reduce process features are more easily reflected in the textual form of data under the MapReduce model, this paper performs Wordcount code transfer through SFTP protocol using self-made data of different textual amounts and two world-famous kinds of literature.

This paper conducted Wordcount tests on the same dataset in the same environment with only Hadoop and other architectures enabled as mentioned above, where the dataset used five grades of data, of which self-created data samples of 10, 1000, and 1w words, and used Jane Eyre (about 20w words) and Shakespeare anthology (about 80w words) as large-scale data samples for the tests, and took ten times the average value as the final experimental conclusion, the following are the experimental results as shown in Figs. 1, 2, 3, 4.

For the Wordcount program, only the Hadoop test used the Wordcount code sample from the Hadoop installation package MapReduce/Hadoop-MapReduce-examples-2.7.7.jar, and the other test used the idea of spark and hive databases to write a

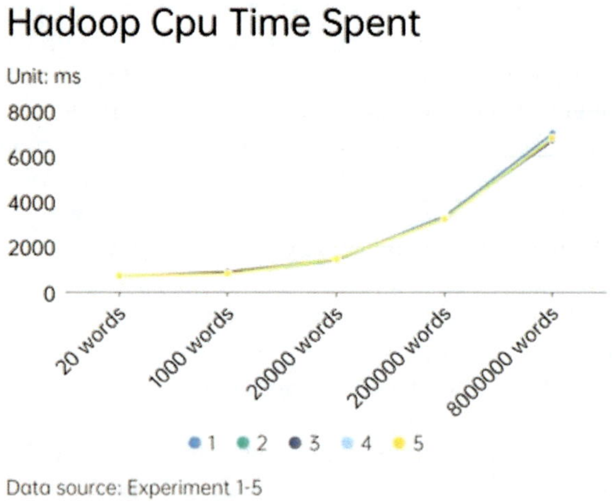

Fig. 1 Hadoop CPU time spent (*Photo credit* Original)

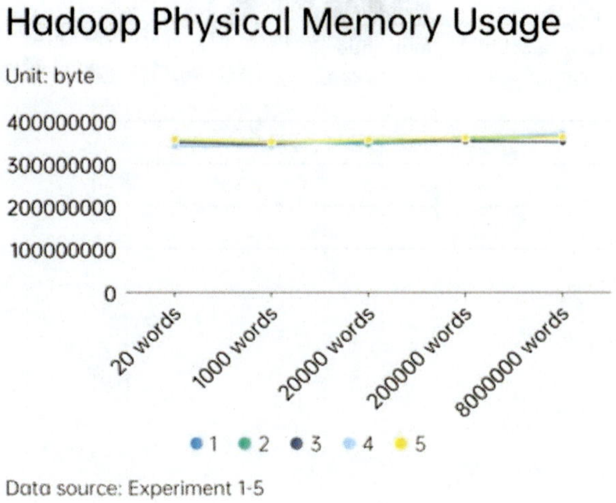

Fig. 2 Hadoop physical memory usage (*Photo credit* Original)

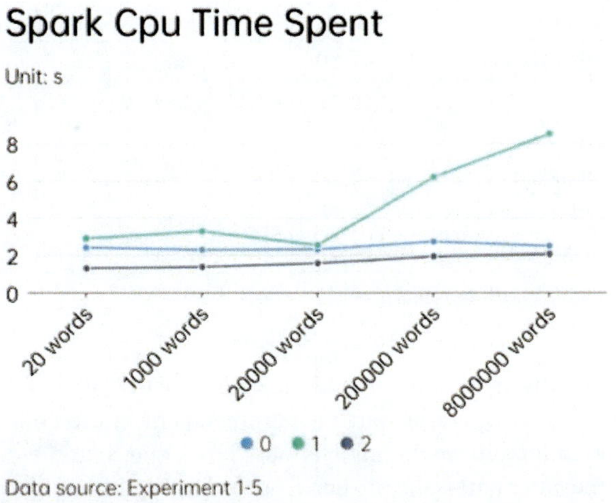

Fig. 3 Spark CPU time spent (*Photo credit* Original)

Wordcount program with Wordcount function for testing. The following are the experimental results of the second test, where the maximum memory usage of the architecture itself is 74.5 kb, where job0 is specifically the time to load data from the hive database to spark, job1 data sample overall MapReduce time, and job2 is the time to write the results.

Tables 1 and 2 compare the data of two tests, in which the time of the test using Spark only shows the time of MapReduce, the program pre-compilation. Other time references are in Table 1. So unexpectedly, this paper used Spark. Hive database test time that only takes MapReduce is also more significant than the test that only uses Hadoop. For the convenience of expression, this paper will not Spark an experiment called experiment one. The Spark experiment is called experiment two. From the

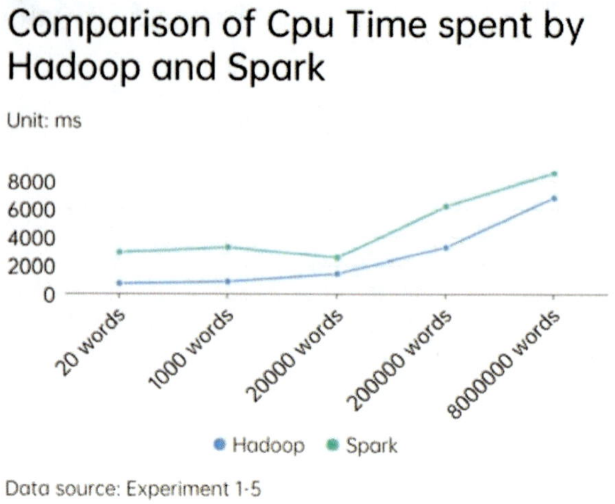

Fig. 4 Comparison of CPU time spent by Hadoop and Spark (*Photo credit* Original)

Table 1 Comparison of CPU time spent in different jobs by Hadoop with Spark

Text volume	Job0	Job1	Job2	Total time
20	2.38s	2.87s	1.3s	6.55s
1000	2.26s	3.25s	1.37s	6.88s
20,000	2.28s	2.51s	1.56s	6.35s
200,000	2.7s	6.17s	1.92s	10.79s
8,000,000	2.47s	8.5s	2.04s	13.01s

(*Table credit* Original)

growth trend of the fit and the data statistics of experiment two, this paper can easily find that Spark as part of the time spent in the compilation process and the whole architecture (including Hadoop, Hive, and Spark, etc.) link communication process, the latter part is the big head; accordingly, although experiment two allocates 434 MB of space, the maximum memory usage is stable at 74.5 KB, while the memory usage of Experiment 1 is extensive and increases with data. Test two is real-time combined with the Hive database and other tasks, so it saves a lot of space but consumes more time. The program has a more significant advantage for distributed arrays with mainly small-capacity nodes, and the time loss is within the acceptable range. The combination of Spark and Hive is a widespread use of the program. The paper would be two way communication links and other directions. The combination of Spark and Hive is a trendy solution. This paper can optimize the communication links between the two to get an excellent solution that saves space and reduces time.

Table 2 Experimental results

Text volume (words)	Time (ms)	Output (words)	RAM	Spark
20	692	46	348 MB	Not used
20	2873	46	74.5 KB	Used
1000	842	4912	347 MB	Not used
1000	3252	4912	74.5 KB	Used
20,000	1404	56,350	352 MB	Not used
20,000	2511	56,350	74.5 KB	Used
20w	3258	321,470	357 MB	Not used
20w	6173	321,470	74.5 KB	Used
80w	6782	535,405	361 MB	Not used
80w	8542	535,405	74.5 KB	Used

(*Table credit* Original)

4 Conclusion

This paper uses the MapReduce model for Wordcount jobs. By inputting different sizes of text tasks to two different frameworks of Hadoop and Spark, recording the detailed data of each task execution, and analyzing its mean value, this experiment not only the unique characteristics of Hadoop and Spark in Wordcount operations using the MapReduce model, but also the innovative ideas that can be optimized and improved in the future. After comparing only Hadoop with the experiments combining Scala and Spark, Spark occupies more space. In the production environment, due to memory limitation, the execution of Job may fail due to insufficient memory resources, and Hadoop still occupies an important position in serverless computing. Reduce stage, for complex computation, it is necessary to use multiple Mrs, which involves drop disk and disk IO, and is inefficient; while in Spark, a Job can contain multiple RDD (Resilient Distributed Data) conversion operators, and multiple Stages can be generated during scheduling to achieve more complex functions. Hadoop was initially designed for one-time data computation, while Spark is an optimized computation process based on traditional MapReduce. Hadoop is suitable for static data, while Spark is suitable for streaming and iterative data. Hadoop is computed at the disk level, while Spark is computed in memory. Hadoop writes data to disk after each processing, and this approach has an inherent advantage in dealing with errors that occur in the system. Spark data objects are stored in Resilient Distributed Data sets that may exist in memory and on disk. Additionally, RDDs have full disaster recovery capabilities.

References

Castro D, Kothuri P, Mrowczynski P et al (2019) Apache Spark usage and deployment models for scientific computing. In: EPJ web of conferences. vol 214. EDP Sciences, p 07020

Dean J, Ghemawat S (2008) MapReduce: simplified data processing on large clusters. Commun ACM 51(1):107–113

Durve VR, Mhatre VM (2015) Comparative study of secure file transfer mechanisms: File transfer protocol over SSL (FTPS), secure shell file transfer protocol (SFTP). Int J Sci Res Develop 3(4):2077–2080

Giménez-Alventosa V, Moltó G, Caballer M (2019) A framework and a performance assessment for serverless MapReduce on AWS Lambda. Futur Gener Comput Syst 97:259–274

Grossi M, Crippa L, Aita A et al (2021) A serverless cloud integration for quantum computing. arXiv preprint arXiv:2107.02007.

Kuśnierz J, Padulano VE, Malawski M et al (2022) A serverless engine for high energy physics distributed analysis. In: 2022 22nd IEEE international symposium on cluster, cloud and internet computing (CCGrid). IEEE, pp 575–584

Shvachko K, Kuang H, Radia S, Chansler R (2010) The Hadoop distributed file system. In: 2010 IEEE 26th symposium on mass storage systems and technologies (MSST). IEEE, pp 1–10

Thain D, Tannenbaum T, Livny M (2005) Distributed computing in practice: the condor experience. Concurrency Comput Pract Exp 17(2–4):323–356

Vavilapalli VK, Murthy AC, Douglas C et al (2013) Apache Hadoop yarn: yet another resource negotiator. In: Proceedings of the 4th annual symposium on cloud computing, pp 1–16

Zaharia M, Xin RS, Wendell et al (2016) Apache spark: a unified engine for big data processing. Commun ACM 59(11):56–65

The Advance and Performance Analysis of MapReduce

Rongpei Han and Yiting Wang

Abstract Cloud computing is highly praised for its high data reliability, lower cost, and nearly unlimited storage. In cloud computing projects, the MapReduce distributed computing model is prevalent. MapReduce distributed computing model is mainly divided into the Map and Reduce functions. As a mapper, the Map function is responsible for dividing tasks (such as uploaded files) into multiple small tasks executed separately; As a reducer, the Reduce function is responsible for summarizing the processing results of multiple tasks after decomposition. It is a scalable and fault-tolerant data processing tool that can process huge voluminous data in parallel with many low-end computing nodes. This paper implements the wordcount program based on the MapReduce framework and uses different dividing methods and data sizes to test the program. The common faults faced by the MapReduce framework also emerged during the experiment. This paper proposes schemes to improve the efficiency of the MapReduce framework. Finally, building an index or using a machine learning model to alleviate data skew is proposed to improve program efficiency. The application system is recommended to be a hybrid system with different modules to process variant tasks.

1 Introduction

Cloud computing is highly praised for its high data reliability, lower cost, and nearly unlimited storage. Cloud computing takes the technology, services, and applications like the Internet and turns them into a self-service utility (Sosinsky 2011). In other

Rongpei Han and Yiting Wang these authors contributed equally.

R. Han (✉)
FedUni Information Engineering Institute, Hebei University of Science and Technology, Shijiazhuang 050000, China
e-mail: rhan@students.federation.edu.au

Y. Wang
Chengdu University of Technology, Oxford Brookes University, Chengdu 610000, China

S. Yadav et al. (eds.), *Proceedings of 2nd International Conference on Artificial Intelligence, Robotics, and Communication*, Lecture Notes in Electrical Engineering 1063, https://doi.org/10.1007/978-981-99-4554-2_20

words, a network that offers resources is what cloud computing is. The "cloud" allows users to access resources at any time and use them as needed, with the potential for limitless growth. For example, the "cloud" is comparable to a natural gas company. People can open the stove anytime, and there is no limit to the amount of gas. Today, cloud computing is highly praised for its high data reliability, low cost, and almost unlimited storage space. Cloud technology has gone deep into storage, medical care, finance, education, and other fields. For example, the primary functions of cloud storage are data management and storage, providing consumers with storage, backup, recording, and other services, which significantly facilitates users' information management. In addition, large Internet companies known to the public, such as Google and Baidu, have cloud storage services. It can be seen from this that people can avoid spending money on creating data centers and buying hardware and software by applying cloud computing. Because of its inherent advantages, cloud computing is bound to grow in various fields.

In cloud computing projects, the MapReduce distributed computing model is prevalent. Over four years in the early twentieth century, Google internal MapReduce programs were built, and each day, on average, 100,000 MapReduce jobs are run on Google's clusters, processing more than 20 petabytes of data (Dean and Ghemawat 2004). And MapReduce distributed computing model is mainly divided into the Map and Reduce functions. As a mapper, the Map function is responsible for dividing tasks (such as uploaded files) into multiple small tasks executed separately; as a reducer, the Reduce function is responsible for summarizing the processing results of multiple tasks after decomposition. It is a scalable and fault-tolerant data processing tool that can process huge voluminous data in parallel with many low-end computing nodes (DeWitt and Stonebraker 2008).

This paper presents the deployment of the wordcount application designed with MapReduce as the framework in Tencent Cloud. It can be used to count the frequency of words in documents. At the same time, we will also discuss how to pre-process and segment files to achieve better program efficiency. To put it simply, after the program is deployed, the .TXT format text uploaded to Tencent Cloud Bucket 1 will trigger the trigger of Bucket 1 to execute the Map function, split the uploaded .TXT format text into multiple small texts and process them and store them in Bucket 2. Finally, Bucket 2 will trigger the reduce function to summarize each result. Although most documents can be processed through this process, considering uploading large documents and optimizing programs, data pre-processing will improve the program's efficiency and reduce the time spent running the program. Therefore, in the experiment, we tried to test the processing of splitting the original document into two, three, and four parts by changing the Map function. Still, the test results were insufficient to find a suitable optimization scheme. And in the experiment, this paper finds that simply increasing the segmentation times cannot effectively shorten the program's running time. Finally, this paper discusses the solution about pre-process the data by adding B + tree retrieval, hierarchical clustering, unsupervised classification, and other machine learning technologies.

2 Methods

This paper uses Tencent Cloud's serverless platform to run the MapReduce-based wordcount program. This paper chooses to run this program on the cloud platform rather than on a single computer is that the cloud platform has more powerful computing power than a single computer. The number of processes managed by a single computer operating system is usually a hundred. In contrast, the number of processes managed by the operating system of the cloud computing center can reach millions. Moreover, many enterprises are transforming to be intelligent, so cloud computing is more suitable for the future development trend. MapReduce is a distributed computing framework that decomposes large data operation jobs into a single task that can be executed parallel across server clusters. It is suitable for large-scale data processing. If it is deployed to the local machine for processing, the hardware requirements are high, and the processing time is long. From this point of view, it is hoped that using a cloud computing platform to build MapReduce frame and process data will give full play to both advantages, making the program more efficient.

2.1 MapReduce

MapReduce is a framework that is used to divide big data into sets parallel and process each set of data simultaneously. The first time it was proposed by two Google researchers. Dean and Ghemawat et al. designed the component of MapReduce composed of two parts, including the Map and Reduce function and the work of MapReduce based on key/value pairs (Dean and Ghemawat 2004). The function of Map processes input content to generate several key/value pairs, meaning it divides a complex task into several simple tasks to reduce the scale of datasets, and these simple tasks independently compute. The function of Reduce is an aggregate process. Reduce receives the computing result from Map (Sardar and Ansari 2018). After, the specific function of Reduce can be customized by users to merge all the same values from Map.

2.2 Serverless Computing

Serverless computing is a cloud-native development model in which developers do not construct and control any infrastructure (Baldini et al. 2017). Developers focus on constructing and running code instead of managing servers. Many large cloud vendors have already developed serverless computing services, such as Amazon, Alibaba, Tencent, Google, etc. The user can deploy code directly to the production environment today, with cloud computing providers managing the physical servers and dynamically assigning resources on their behalf (Giménez-Alventosa et al. 2019). In this paper, two main services will be used to complete MapReduce. The first is a bucket (an object-based data store) which requires users to store tons of datasets, with loads of users Remotely accessing data. Another one is Serverless Cloud Function (SCF) which provides the execution environment for developers. Event-Triggered included in SCF is a serverless operation method. It can be decided when the SCF is triggered.

2.3 Wordcount

Considering the traits of MapReduce, Wordcount is an exciting and simple example to express the computing process. Therefore, Wordcount used MapReduce was applied on Tencent Cloud using Python. There are characteristics of Wordcount shown below: (1) The input file is allowed to upload to a starting bucket to store the original text and trigger the Map function. (2) Map function splits the uploaded file and processes the file into two files according to alphabetical order, which means each word will be read in turn, and it is determined whether the word's initial is in a-m/A-M. If the word's initial is in this range, this word will be written in the first file and append "1" behind this word. Otherwise, this word will be written in the second file as well as append "1" behind this word. (3) Two files can be simultaneously processed to generate key/value pairs. Once the words are processed, the middle bucket will store two files completed by the Map function and trigger the Reduce function. (4) Reduce function processed the two files simultaneously from the middle bucket. A dictionary is defined in advance to store each key and corresponding value. After, the main task for Reduce function is calculating the frequency of occurrence of each word, which means once the same key is read, the value will be added. In the end, reduce function generates new key/value pairs—frequency statistics for words with initials a-m/A-M in the first file and other words in the second file.

The general process of program operation is as follows: when the text is input to Cloud Object Storage (COS) Bucket 1, Bucket 1's trigger will trigger the Mapper function, and the intermediate text results processed by the Mapper function will be stored in Bucket 2. Bucket 2 trigger will trigger the reducer function. Finally, the results processed by the reducer function are stored in Bucket 3. The final result will be written to Bucket 3, as shown in Figs. 1 and 2.

In the process of using sample test data, this paper proposes two aspects that can be optimized. The first is to build a hybrid system to pre-process data according to different sample sizes. The second is the remission for data skew.

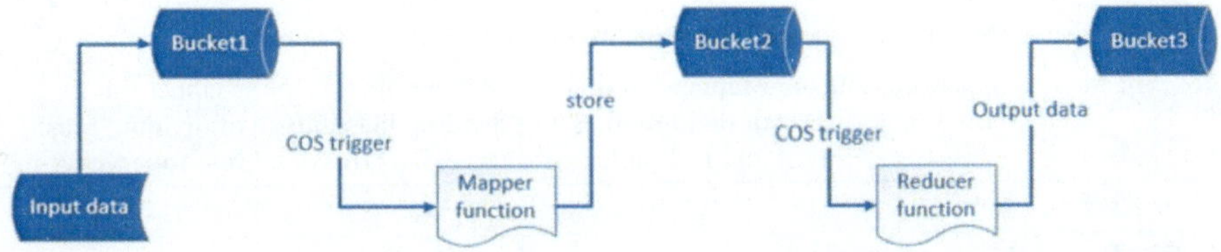

Fig. 1 Flowchart of wordcount program operation

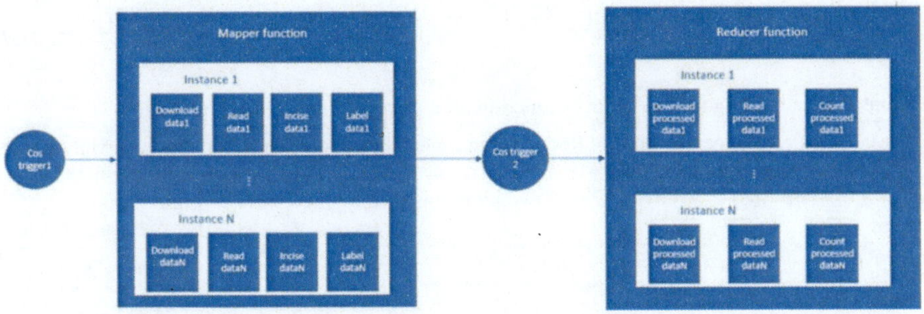

Fig. 2 Mapper function and reducer function

3 Experimental Results and Analysis

This paper changes the text size without changing the segmentation method of words read from the text. The experimental conclusion is that the mapper and reducer's runtime increase with the text's increase. However, the running memory of the mapper and reducer does not change significantly with the change in text size, as shown in Table 1.

Then, this paper changes the way of processing the text content. When reading the text, this paper divides the text content into two, three, and four parts according to the words' initial order, trying to find the appropriate segmentation method for text with different sizes, as indicated in Tables 2 and 3.

After testing, there is a big gap between the experimental expectation and the results. This paper argues that first, the text of the test data is too small to reflect the

Table 1 The same function handles different tasks

The size of files	Mapper run time (ms)	Mapper run memory (MB)	Reducer run time (ms)	Reducer run memory (MB)
14.47 KB	870	18.49	316	18.32
3.63 MB	1370	18.58	1296	18.65
4.14 MB	1477	18.59	1010	18.25
9.01 MB	2731	18.34	1641	18.13

Table 2 The running time of mapper function after changing times of segmentation

Size of files	Mapper run time of original code (ms)	Mapper run time of dividing files into two parts (ms)	Mapper run time of dividing files into three parts (ms)	Mapper run time of dividing files into four parts (ms)
4.01 KB	269	432	444	514
32.5 KB	263	1,750	1,843	1,846
1.2 MB	595	51,584	50,502	54,303
3.62 MB	1156	140,326	154,505	154,578
7.25 MB	1708	312,212	307,489	313,861

Table 3 The running time of reducer function after changing times of segmentation

Size of files	Reducer run time of original code (ms)	Reducer run time of dividing files into two parts (ms)	Reducer run time of dividing files into three parts (ms)	Reducer run time of dividing files into four parts (ms)
4.01 KB	248	670	689	1080
32.5 KB	294	451	846	1,168
1.2 MB	549	1,026	966	3,312
3.62 MB	992	1,621	1,746	2,245
7.25 MB	1,710	3,188	2,766	5,278

segmentation method's advantages. Second, the program itself may have limitations, resulting in low efficiency.

4 Discussion

4.1 Indexing

This paper suggests pre-processing the data before running MapReduce to improve the program's efficiency. Different methods are adopted to deal with different sample sizes and segmentation difficulties. For the samples with large scale and easy segmentation, the samples are first segmented and indexed. The traditional relational database establishes a B + tree index or a B-tree index when building an index. Only the indexed keys and a pointer to the corresponding record are kept in the leaf nodes of a B + tree (Barranco et al. 2008). In the B + tree index, data is arranged in order, which makes searching extremely simple. Because the data of the B-tree index is scattered in each node, it is not easy to achieve this. Therefore, for the samples such as wordcount, which is easy to classify data, you can use a non-clustered index to build an index. The purpose of creating an index is not to query specific data. This structure can be used quickly and does not require more space to store specific data because the specific data in the sample does not need to be stored in the leaf node. Although this traditional database indexing method is not the only solution, it can still be used as a reference to provide some solutions for today's technical problems. For this structure, the specific operations are as follows: First, according to the user's needs, customize the indexing standards and split them. Secondly, each segmented sample corresponds to a unique value as the key value to build a B + tree index. Finally, establish a B + tree index and store the corresponding key values. For example, before wordcount runs, we customize a B + tree index containing the letters a–z. After running the mapper function, attach the key value corresponding to the B + tree to each word, and put the words with the same key value into the same file for Reduce operation. In this way, you can directly search the index and read the results when searching the related content. For samples with large scale but not accessible to segment, an unsupervised classification machine learning method can be added to complete the segmentation by using the information of the sample itself. Then the samples can be processed in the same way for samples with large scale and easy to segment as the above. In most scenarios, especially when processing big data tasks, such methods will be very efficient and appropriate. However, this paper also finds other problems: when dealing with small samples, the time and memory cost of indexing and classifying small samples using unsupervised learning is far greater than that of directly conducting MapReduce operations. Building indexing takes much more time than directly processing small samples. Meanwhile, when building a B + tree index, the space will be occupied additionally. Therefore, it is better to deal with small texts directly. Based on different processing methods for large and small samples, the final system should be a hybrid system with different modules to process various tasks.

4.2 Processing Data Skew

In addition to data pre-processing, this paper notices the problem of data skew in the experiment. Physical characteristics of objects (such as normally distributed heights of people) and hot spots on specific subsets of a whole domain frequently create data skew (e.g., word frequency in documents following a Zipf distribution) (Chen et al. 2015). When the nodes do not reach the full load state, there will be a data skew problem, leading to low program efficiency. The challenge is for unbalanced loads that are distributed to reducers (Irandoost et al. 2019). However, many samples are different in size, and cutting the samples is easy to cause information loss. Therefore, it isn't easy to ensure that all nodes reach the full load state in this case, and data skew is challenging to solve completely. But the data skew problem cannot be ignored, and it is a direction that can optimize the program. To alleviate the data skew problem, the user tries to increase the parallelism of the shuffle and adjust it according to the situation of the server. When data skew occurs, the user can appropriately increase the parallelism and assign different keys originally assigned to the same task to other tasks. In addition, the operating system's fragmented space also causes data skew to a certain extent. For this, the operating system sets a cache area. The file system or the operating system will cache the data to be written in memory through the write cache area. After some time, the data will be continuously written to the disk, reducing many direct writes of small-size data.

However, today's distributed computing frameworks will undertake some more critical tasks. AI applications, such as automatic driving, medical image diagnosis, and intelligent speech recognition, must process many images and speech, and their data size and complexity are far greater than text. For example, many high-level applications require the preliminary task of semantic segmentation (Benois-Pineau and Zemmari 2021). But due to natural language limitations, the data of speech type is difficult for each voice information to be consistent or similar in size, making it difficult to divide and evenly distribute to each node. If the data is cut directly and processed in a multi-process manner, it may lead to data loss. For example, in the aspect of image segmentation, although people have proposed different segmentation methods, when they are applied to a specific type of image, due to the limitations of each segmentation method, they still need to combine each segmentation method to get a better segmentation result. Not only the application of AI but also some mathematical and physical problems, such as partial differential equations, need to be solved with the help of deep learning models or statistical models. However, when the computing process is deployed to supercomputers, the data skew problem cannot be avoided. The deep learning model uses the gradient descent method to solve problems. When multiple nodes gather gradient information, the data imbalance will cause the gradient information of each computing node to be inconsistent in the amount of data, and each node cannot operate with a full load and high efficiency. Meanwhile, combining results among multiple computing nodes may also lead to errors in the results. This is the reason why the problem is challenging to solve.

5 Conclusion

This paper tested the wordcount program based on the MapReduce framework and obtained the corresponding experimental data such as running time. Compared with other experiments, this paper compares the time and space cost of the program by using the Wordcount instance to modify the segmentation method in the mapper function. According to the results of the experiment, it is concluded that only increasing the segmentation times cannot effectively improve the program's efficiency. In this regard, this paper believes that data pre-processing can effectively improve the efficiency of functions and proposes some methods, such as building an index through traditional B + trees or mitigating data skew through machine learning techniques to achieve the purpose of pre-processing and improving program efficiency. This research aims to provide methods that can improve the efficiency of functions in data pre-processing by integrating the above technologies. Further studies, this paper will continue to study solutions for data volume inconsistency of gradient information of each computing node caused by data imbalance and the combination of every node's results in deep learning.

References

Baldini I, Castro P, Chang K et al (2017) Serverless computing: current trends and open problems. In: Research advances in cloud computing. Springer Singapore, pp 1–20. https://doi.org/10.1007/978-981-10-5026-8_1

Barranco CD, Campaña JR, Medina JM (2008) A B +—tree based indexing technique for fuzzy numerical data. Fuzzy Sets Syst 159(12):1431–1449. https://doi.org/10.1016/j.fss.2008.01.006

Benois-Pineau J, Zemmari A (2021) Multi-faceted deep learning. Springer International Publishing

Chen Q, Yao J, Xiao Z (2015) LIBRA: lightweight data skew mitigation in MapReduce. IEEE Trans Parallel Distrib Syst 26:2520–2533

Dean J, Ghemawat S (2004) MapReduce: simplified data processing on large clusters. In: Proceedings of the 6th conference on symposium on operating systems design & implementation, vol 6

DeWitt D, Stonebraker M (2008) MapReduce: a major step backwards. Database Column 1:23

Giménez-Alventosa V, Moltó G, Caballer M (2019) A framework and a performance assessment for serverless MapReduce on AWS Lambda. Future Gener Comput Syst 97:259–274. https://doi.org/10.1016/j.future.2019.02.057

Irandoost MA, Rahmani AM, Setayeshi S (2019) A novel algorithm for handling reducer side data skew in MapReduce based on a learning automata game. Inform Sci Int J 501:501

Sardar TH, Ansari Z (2018) Partition based clustering of large datasets using MapReduce framework: an analysis of recent themes and directions. Future Comput Inform J 3(2):247–261. https://doi.org/10.1016/j.fcij.2018.06.002

Sosinsky BA (2011) Cloud computing bible. Wiley Pub

An Efficient Model for Dorsal Hand Vein Recognition Based on Combination of Squeeze-and-Excitation Block and Vanilla ResNet

Cai Zong, Peirui Bai, Qingyi Liu, Zheng Li, Xiaoxiao Ge, Rui Yang, Tao Xu, and Guang Yang

Abstract The dorsal hand vein (DHV) biometrics is commonly employed in personal verification or identification due to its excellent anti-counterfeit and liveness detection capabilities. In the field of DHV biometrics, the performance of deep convolutional neural networks (DCNNs) is limited due to insufficient labeled data, less discriminative features, and various image qualities of different dataset. In this paper, we take the vanilla ResNet50 as backbone, combine the Squeeze-and-Excitation block and adopt knowledge transfer to enhance the recognition performance. First, SE-ResNet50 model is constructed by embedding the Squeeze-and-Excitation module into each convolutional block of the vanilla ResNet50. Second, the transfer learning strategy is adopted to speed up training efficiency and enhance generalization capability. That is, the MPD palmprint dataset is employed to pre-train the SE-ResNet model. The parameters of the pre-trained model are adjusted using DHV datasets to achieve knowledge transfer. Three datasets were conducted in our experiments. The experimental results demonstrated that the adoption of attention mechanism and transfer learning can significantly improve the recognition accuracy. The proposed SE-ResNet50 model achieves competitive performance of the state of the art with higher computational efficiency and generalization capability.

1 Introduction

Dorsal hand vein (DHV) biometrics is a promising technology for personal verification or identification. It has drawn much attention in recent years (Uhl et al. 2020; Jia et al. 2021). The vein patterns of the DHV are unique and stable. It has good anti-counterfeit and liveness detection capability. The early DHV recognition methods depend mainly on hand-crafted features, e.g., the texture-based descriptors

C. Zong · P. Bai · Q. Liu · Z. Li · X. Ge · R. Yang · T. Xu · G. Yang (✉)
College of Electronic and Information Engineering, Shandong University of Science and Technology, Qingdao, China
e-mail: gyang@sdust.edu.cn

S. Yadav et al. (eds.), *Proceedings of 2nd International Conference on Artificial Intelligence, Robotics, and Communication*, Lecture Notes in Electrical Engineering 1063, https://doi.org/10.1007/978-981-99-4554-2_21

(Vairavel et al. 2019), the structure-based descriptors (Kauba and Uhl 2018; Guo et al. 2020), the subspace learning-based descriptors (Zhang et al. 2016), and the direction coding-based descriptors (Lee et al. 2016). More recently, deep convolutional neural networks (DCNNs) have been introduced into the field of DHV biometrics (Li et al. 2016). However, there exist severe challenges, such as the lack of large-scale trainable data and no guarantee of data quality (Ribaric et al. 2016). Fortunately, transfer learning provides an effective paradigm for adapting machine learning methods to new application domains. To better leverage the well-trained models, fine-tuning, knowledge distillation or domain adaptation can be applied to realize knowledge transfer in intra-domain or cross-domain (Liu et al. 2112). Several pioneering attempts on transfer learning in DHV biometrics have been explored. For example, Wang et al. (2017) proposed a transfer learning model to train task-specific models using a coarse-to-fine scheme. The model achieved satisfactory results by effectively applying the DCNN model to a small-scale DHV recognition. Using transfer learning and DenseNet-161, Kuzu et al. (2020) designed a highly effective vein-based biometric recognition system. Gu et al. (2021) proposed a technique that utilized transfer learning to integrate characteristics of local binary patterns into the ResNet50 architecture. Rajalakshmi and Annapurani (2022) suggested using feature learning and transfer learning techniques to identify the palmar vein. All of the above methods using transfer learning have achieved acceptable performance in DHV recognition. To improve a network's representational quality by modeling channel-wise relationships, Hu et al. (2018) proposed a Squeeze-and-Excitation Network (SENet), which learns to selectively highlight useful features and reduce less valuable ones employing global information. Joshua et al. (2021) constructed a unique SE-CNN using SENet. The successful results confirmed the technique's reliability for classifying lung nodules. According to the aforementioned studies, the SE module can boost performance to a certain extent.

In this paper, a novel model called SE-ResNet50 is proposed for dorsal hand vein recognition, which aims to comprehensively utilize the advantages of attention mechanism and knowledge transfer. Firstly, a large-scale palmprint dataset is used to train a ResNet50 model combining attention modules. The model can capture spatial relations between different feature maps by the convolutional layers, which integrate the Squeeze-and-Excitation block into the vanilla structures. Secondly, a parameter-tuned transfer learning approach is proposed. By taking advantage of the knowledge learned from the palmprint dataset, the parameters of the SE-ResNet50 model are adjusted to improve the recognition performance of DHV patterns. The main work contributions are as follows: (1) A novel model called SE-ResNet is proposed which outperforms the other three models for DHV recognition. (2) The pre-trained model of palmprint is proposed to improve the DHV recognition performance in our model, which can achieve higher accuracy in less training time.

2 Related Work

2.1 ResNet Network

Being a new popular DCNN model, ResNet (He et al. 2016) solved the vanishing gradient problem by introducing the residual unit. The advantage is the residual structure which solves the problem of deep network degradation, and makes it possible to train deeper neural networks. Xie et al. (2017) further proposed ResNeXt to obtain stronger representation capabilities without increasing the number of parameters. ResNet has been extensively used in numerous image classification tasks as a frequently used deep model.

2.2 *Attention Mechanism of the Squeeze-and-Excitation Block*

The attention mechanism in deep learning is similar to that of human vision (Mnih et al. 2014). That is, the primary attention is preferentially paid on the critical parts among many information pieces, and a small amount of attention is paid on the secondary parts. In 2017, Hu et al. (2018) designed the "Squeeze-and-Excitation network." The SENet won the ImageNet Large Scale Visual Recognition Competition 2017. With the help of the "Squeeze-and-Excitation (SE) block," the Top-5 error rate can be reduced to 2.25%. The structure of SE module is shown in Fig. 1. Roy et al. (2018) proposed three variants of the SE module and demonstrated through experiments that such modules can enhance meaningful features and suppress useless features. To boost the system performance, various attention techniques are presently incorporated into deep models.

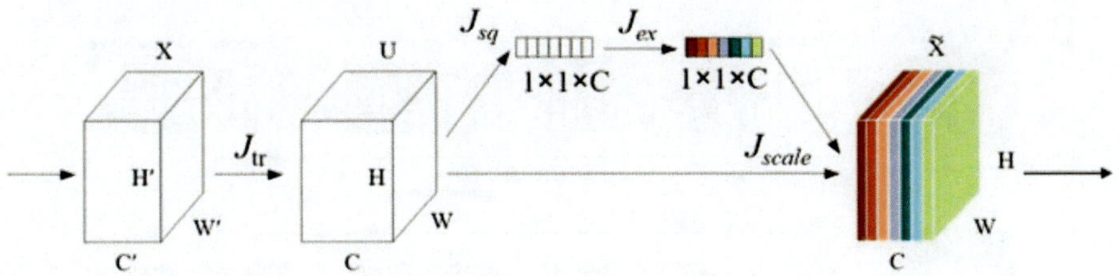

Fig. 1 The structure of SE module in SENet

3 The Proposed SE-ResNet50 Model

3.1 The Model Architecture

First, the interconnected convolutional layers of the vanilla ResNet are adopted as a single block named BasicBlock. The SE-ResNet50 consists of five stages and an output layer. The Stage0 has a convolutional layer with the 1×1 size of kernel and a max pooling layer. Other Stages contain several BasicBlocks followed by an embedded SENet. The output layer has a global average pooling layer and a fully connected layer. Finally, we use the Softmax to output the classification results. The architecture of the SE-ResNet50 model is shown in Fig. 2. In the proposed SE-ResNet50 architecture, ResNet50 extracts feature maps $U \in R^{H \times W \times C}$ from $X \in R^{H' \times W' \times C'}$ by convolution J_{tr}. When generating feature maps, the network assigns the same weight to each channel. Then they are fed into the SE module for receiving more channel-wise information. SE module aims to change the channel weighting through the incorporation of a content-aware mechanism into its structure. It requires a single parameter and a linear scalar to each channel to be assigned in its basic form and uses squeeze function and excitation function to selectively highlight useful features and reduce less valuable ones.

Squeeze function J_{sq} conducts average pooling operation for each individual channel, and produces a channel descriptor z, which can generate by compressing the spatial dimension $W \times H$ of U, such that the c-th element of J_{sq} is calculated by

$$z = J_{sq}(U_c) = \frac{1}{H \times W} \sum_{i=1}^{H} \sum_{j=1}^{W} U_c(i, j) \tag{1}$$

Excitation function J_{ex} consists of two fully connected layers and a ReLU nonlinear layer to get channel-wise dependencies, which can be calculated by

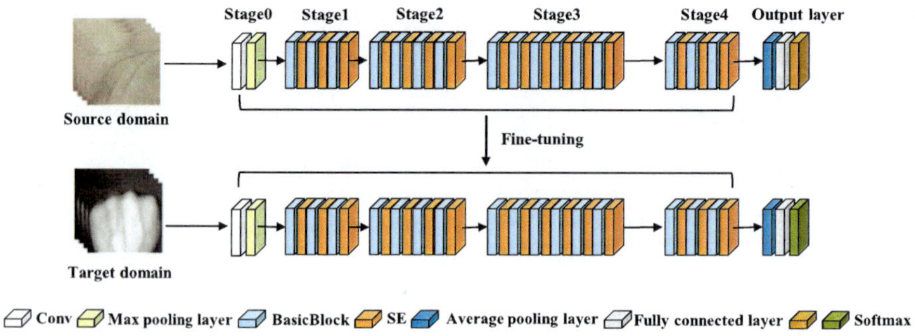

Fig. 2 The architecture of SE-ResNet50 model and the process of knowledge transfer

$$J_{\text{ex}}(z, f) = s(g(z, f)) = s(f_2\delta(f_1 z)) \tag{2}$$

where δ, s refers to the ReLU and Sigmoid function, and f_1, f_2 are fully connected layers. After that, return to the channel dimension of the transformation output U. The final output is obtained by J_{scale}:

$$\tilde{x}_c = J_{\text{scale}}(U_c, J_{\text{ex}}) = J_{\text{ex}} U_c \tag{3}$$

$$\tilde{X} = [\tilde{x}_1, \tilde{x}_2, \ldots, \tilde{x}_C] \tag{4}$$

By using channel adjustment for each convolutional block, the weights of each feature map can be fine-tuned. By using the convolutional filters, hierarchical information from the images can be extracted by the SE-ResNet50. Lower layers can only identify borders or extremely high frequencies, whereas higher levels can recognize complicated geometric shapes. The feature vector of the last layer is passed through the average pooling layer and the fully connected layer. Finally, Softmax classifier is used to output the recognition results. Softmax can be expressed as follows:

$$S(y_i) = \frac{\exp(y_i)}{\sum_{j=1}^{C} \exp(y_j)} \tag{5}$$

where C, y_i, and exp(.) stand for the number of categories, probability of class i, and exponential operation.

3.2 *The Solution of Knowledge Transfer*

Compared to the acquisition of DHV images, palmprint images are more convenient to be captured as it does not require professional imaging devices. In addition, the vein-like textures of the palmprint images are similar to that in DHV images. Therefore, it is feasible to achieve knowledge transfer from palmprint images to the DHV images.

In order to realize the knowledge transfer, we take the learning of palmprint images at larger scales as the source learning task, and the learning of DHV images at smaller scales as the target learning task. We use the source learning task to improve the ability of the target learning task by transfer learning. The SE-ResNet50 model firstly is sufficiently pre-trained to effectively extract features of palmprint patterns. Then, the parameters stored in pre-trained model will be used as the initialization of the DHV network. The first five stages of the DHV network are used as a feature extractor, and the output layer uses a new Softmax classifier. Finally, the DHV network is updated with a new classifier through the fine-tuning strategy.

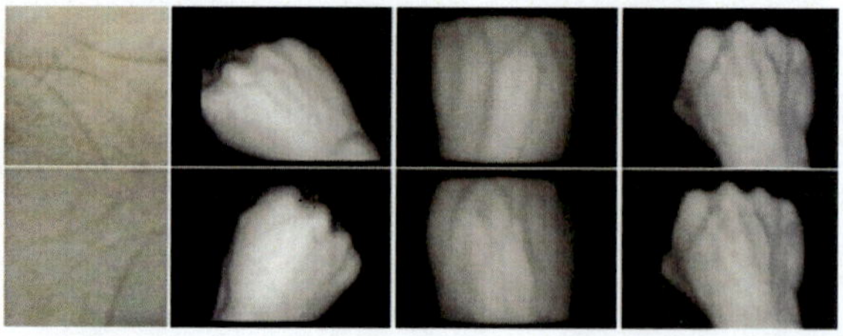

Fig. 3 The example images of **a** MPD, **b** DF, **c** NCUT and **d** JLU

4 Experiments and Discussions

4.1 Datasets and Environment

The MPD dataset, the largest in the field of palmprint identification, contains 16,000 palm images from 200 subjects provided by Tongji University (Zhang et al. 2003). A self-produced dataset called DF-dataset is constructed and contains 3500 DHV images that were collected from 350 volunteers. For each subject, 10 images were acquired for left or right hand. The NCUT dataset contained 2040 reflected images provided by North China University of Technology (Huang et al. 2017). The JLU dataset contained 3680 reflected images provided by Jilin University (Liu et al. 2020). Some example images in those four datasets were shown in Fig. 3. The experimental platform was Intel(R) Xeon(R) Silver 4210R CPU and NVIDIA RTX A6000 GPU with CUDA version 11.1. In the pre-training procedure, we adopted the SGD optimizer and the default values of its parameters such as the initial learning rate, momentum, and weight delay were set to 0.01, 0.9, and 0.01 respectively. Each mini-batch contained 128 images and the training epochs were set to 100. In the fine-tuning procedure, the different settings from pre-training were that the training epochs were set to 75.

4.2 Experimental Results and Analysis

Here, we employed MPD dataset as the source domain for pre-training the four different DCNN models (VGG16 (Simonyan and Zisserman 1409), SE-VGG16, ResNet50, SE-ResNet50) and the three DHV datasets as target domain for fine-tuning that of the models. All three DHV datasets were divided into train set, valid set, and test set in terms of a ratio of 6:2:2. Table 1 listed the recognition rates of DHV recognition without and with MPD pre-training on different DCNN models. We reported the top 1 accuracy on the test set. In fine-tuning procedure, it can be observed that the DHV recognition rates of the proposed SE-ResNet50 model were

Table 1 Comparisons of DHV recognition rates using different DCNN models

Datasets	Methods		Recognition rates (%)			
	Schemes	Epochs	VGG16	SE-VGG16	ResNet50	SE-ResNet50
DF	o.	150	77.52	89.76	91.21	92.42
	w.	75	95.10	95.53	95.00	**98.70**
NCUT	o.	150	93.30	95.12	94.10	94.87
	w.	75	98.71	98.97	99.48	**99.74**
JLU	o.	150	67.93	68.75	74.05	84.10
	w.	75	70.38	82.20	95.65	**95.92**

o./w. note without or with MPD pre-trained model

98.70%, 99.74%, and 95.92% on the DF dataset, NCUT dataset, and JLU dataset respectively, which were higher than the other models.

In order to verify the effectiveness of the MPD pre-trained model for improving the dorsal hand vein recognition. Without using the MPD pre-trained model, we fully trained four different DCNN models on three DHV datasets and the training epoch was set to 150. Experiment results showed that the DHV recognition performance of different models without MPD pre-training was lower than that with MPD pre-training. From the training time and recognition rate of view, the DCNN models with MPD pre-training can achieve higher recognition rate than that of the models without pre-training using only half of the training epochs. In addition, the proposed SE-ResNet50 model can bring about the highest recognition rate compared to the other three models.

4.3 *Evaluation of Knowledge Transfer Between Palmprint and DHV Image Domains*

In several deep learning models that were trained on photos, they learn generic features on the first layer. However, as the network gets deeper, the deeper layers are more focused on learning task-specific features (Yosinski et al. 2014). Therefore, it was needed to evaluate quantitatively the transferability between the two image domains. Here, we adopted the Logarithm of Maximum Evidence (LogME) method (You et al. 2021) to investigate the issue, which was widely applicable to various pre-trained models as well as downstream tasks.

In computer vision, ImageNet1000 is a large-scale dataset containing 1.2 million natural scene images and its pre-trained models have good performance in various downstream tasks. As a result, we employed the ImageNet1000 (Jia et al. 2009) and the MPD as source domain respectively, and exploited three DHV datasets as target domain to assess transfer performance. In fact, the higher the value of LogME score, the better the transfer performance of the pre-trained model. In Table 2, the

Table 2 Assessment of ImageNet and MPD pre-trained models for transfer learning

Experiment settings			
Source	Target	Source model	LogME
ImageNet1000	DF	ResNet50	1.514
MPD	DF	ResNet50	1.515
MPD	DF	SE-ResNet50	**1.515**
ImageNet1000	NCUT	ResNet50	0.903
MPD	NCUT	ResNet50	0.913
MPD	NCUT	SE-ResNet50	**0.915**
ImageNet1000	JLU	ResNet50	1.885
MPD	JLU	ResNet50	1.887
MPD	JLU	SE-ResNet50	**1.888**

combination of MPD and SE-ResNet50 performs the best transfer on the DHV datasets.

4.4 Discrimination of Attention Mechanism-Based Feature Representations

In the traditional machine learning methods, hand-crafted features, e.g., endpoints (Akram et al. 2014), crossing points (Ding et al. 2005), were often employed. Discriminative feature representation has a higher requirement in the field of DHV biometrics. To intuitively disclose the superiority of feature representation by introducing the SE module into the ResNet50 network, Fig. 4 demonstrated the comparisons of Grad-CAM image (Zhou et al. 2016) between the ResNet50 and the SE-ResNet50 network. From the second and third columns, it can be seen that the SE-ResNet50 network captures more hand vein information than the ResNet50 model, such as the intersections in the vein patterns, which was helpful to increase the discriminative capability for personal identification.

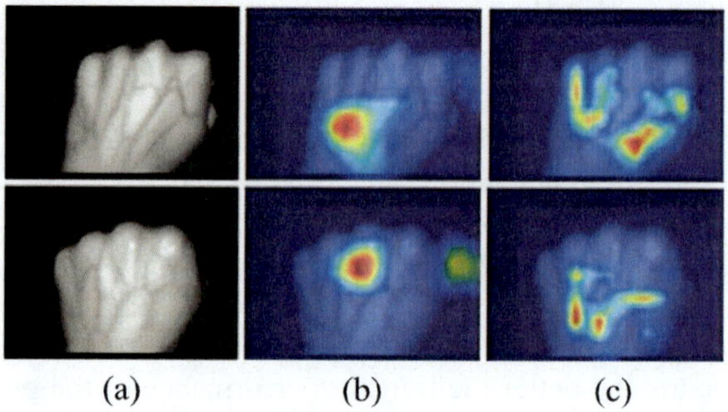

Fig. 4 The first column **a** illustrated two original DHV images. The second column **b** shown the Grad-CAM images of the ResNet50 model. The third column **c** shown the Grad-CAM images of the SE-ResNet50 model

5 Conclusion

In this paper, we proposed a novel model named SE-ResNet50 for DHV recognition. On the basis architecture of ResNet50, attention mechanism implemented by Squeeze-and-Excitation network and knowledge transfer of palmprint domain was realized. The results of confirmatory and comparative experiments demonstrated that the DCNN models with MPD pre-training enhance the recognition rates obviously, and the proposed SE-ResNet50 outperformed several competitive DCNNs. In the next work, we will investigate the effective transfer learning strategies of reducing training parameters while ensuring recognition performance. In addition, we will explore more suitable source domains for knowledge transfer in NIR dorsal hand vein images.

5. References

Akram MU, Awan HM, Khan AA (2014) Dorsal hand veins based person identification. In: 4th international conference on image processing theory, tools and applications (IPTA). IEEE, Istanbul, Turkey, pp 1–6

Ding YH, Zhuang DY, Wang KJ (2005) A study of hand vein recognition method. In: International conference mechatronics and automation. IEEE, Niagara Falls, Canada, pp 2106–2110

Gu GJ, Bai PR, Li H, Liu QY, Han C, Min XL, Ren YD (2021) Dorsal hand vein recognition based on transfer learning with fusion of LBP feature. In: 15th Chinese conference on biometric recognition. Springer, Shanghai, China, pp 221–230

Guo ZY, Ma Y, Min XL, Li H, Liu QY, Han C, Yang G, Bai PR, Ren YD (2020) A novel algorithm of dorsal hand vein image segmentation by integrating matched filter and local binary fitting level set model. In: 7th international conference on information science and control engineering (ICISCE). IEEE, Changsha, China, pp 81–85

He KM, Zhang XY, Ren SQ, Sun J (2016) Deep residual learning for image recognition. In: Proceedings of the IEEE Conference on Computer Vision and pattern recognition. IEEE, pp 770–778

Hu J, Shen L, Sun G (2018) Squeeze-and-excitation networks. In: Proceedings of the IEEE Conference on Computer vision and pattern recognition. IEEE, pp 7132–7141

Huang D, Zhang RK, Yin Y, Wang YD, Wang YH (2017) Local feature approach to dorsal hand vein recognition by centroid-based circular key-point grid and fine-grained matching. Image vis Comput 58:266–277

Jia W, Xia W, Zhang B, Zhao Y, Fei LK, Kang WX, Huang D, Guo GD (2021) A survey on dorsal hand vein biometrics. Pattern Recogn 120:108–122

Jia D, Wei D, Richard S, Li JL, Kai L, Li FF (2009) Imagenet: a large-scale hierarchical image database. In: Proceedings of the IEEE Conference on Computer Vision and pattern recognition. IEEE, pp 248–255

Joshua N, Stephen E, Bhattacharyya D, Chakkravarthy M, Kim HJ (2021) Lung cancer classification using squeeze and excitation convolutional neural networks with grad cam++ class activation function. Traitement du Signal 38(4)

Kauba C, Uhl A (2018) Shedding light on the veins-reflected light or transillumination in hand-vein recognition. In: 11th IAPR international conference on biometrics (ICB). IEEE, Queensland, Australia, pp 283–290

Kuzu RS, Maiorana E, Campisi P (2020) Vein-based biometric verification using transfer learning. In: 43rd international conference on telecommunications and signal processing (TSP). IEEE, Milan, Italy, pp 403–409

Lee JC, Lo TM, Chang CP (2016) Dorsal hand vein recognition based on directional filter bank. SIViP 10(1):145–152

Li XX, Huang D, Wang YH (2016) Comparative study of deep learning methods on dorsal hand vein recognition. In: 10th Chinese conference on biometric recognition. Springer, Tianjin, China, pp 296–306

Liu F, Jiang SK, Kang B, Hou T (2020) A recognition system for partially occluded dorsal hand vein using improved biometric graph matching. IEEE Access 74525–74534

Liu Y, Zhang W, Wang J, Wang JY (2021) Data-free knowledge transfer: a survey. arXiv:2112.15278

Mnih V, Heess N, Graves A (2014) Recurrent models of visual attention. In: Advances in Neural Information Processing Systems, vol 27

Rajalakshmi M, Annapurani K (2022) A deep learning based palmar vein recognition: transfer learning and feature learning approaches. In: Proceedings of international conference on deep learning, computing and intelligence. Springer, Singapore, pp 581–591

Ribaric S, Ariyaeeinia A, Pavesic N (2016) De-identification for privacy protection in multimedia content: a survey. Signal Process Image Commun 47:131–151

Roy AG, Navab N, Wachinger C (2018) Concurrent spatial and channel 'squeeze & excitation' in fully convolutional networks. In: International conference on medical image computing and computer-assisted intervention. Springer, Cham, pp 421–429

Simonyan K, Zisserman A (2014) Very deep convolutional networks for large-scale image recognition. arXiv:1409.1556

Uhl A, Busch C, Marcel S, Veldhuis RNJ (2020) Handbook of vascular biometrics. Springer Nature

Vairavel KS, Ikram N, Mekala S (2019) Performance analysis on feature extraction using dorsal hand vein image. Soft Comput 23(18):8349–8358

Wang J, Wang GQ, Zhou M (2017) Bimodal vein data mining via cross-selected-domain knowledge transfer. IEEE Trans Inf Forensics Secur 13(3):33–744

Xie S, Girshick R, Dollár P, Tu Z, He K (2017) Aggregated residual transformations for deep neural networks. In: Proceedings of the IEEE conference on computer vision and pattern recognition. IEEE, pp 1492–1500

Yosinski J, Clune J, Bengio Y, Lipson H (2014) How transferable are features in deep neural networks? In: Advances in Neural Information Processing Systems, vol 27

You KC, Liu Y, Wang JM, Long MS (2021) Logme: practical assessment of pre-trained models for transfer learning. In: International conference on machine learning. PMLR, pp 12133–12143

Zhang YY, Zhang L, Zhang RX, Li SX, Li JL, Huang FY (2020) Towards palmprint verification on smartphones. arXiv:2003.13266

Zhang D, Guo ZH, Gong YZ (2016) Dorsal hand recognition. In: Multispectral biometrics. Springer, Cham, pp 165–186

Zhou BL, Khosla A, Lapedriza A, Oliva A, Torralba A (2016) Learning deep features for discriminative localization. In: Proceedings of the IEEE Conference on Computer Vision and pattern recognition. IEEE, pp 2921–2929

Design and Implementation of Online Book Sale System

Zhixiong Miao

Abstract Whether it is a large enterprise or a small and medium-sized enterprise, information-based office work is necessary for the current environment. It can effectively improve work efficiency and quickly solve various complex problems. Hover, the existing work lacks the elaboration and explanation of e-books and borrowing. So, in this paper, an online book sales system is designed by using SQL, which includes common book information, order information, and employee information, and adds a module to both the sale and lease of physical and electronic books. In this system, three roles are designed, namely customer, employee, and administrator. Customers can enter the system by registering and logging in to their account to view books and their details and can select the books they want, purchase or borrow books, and select physical or electronic versions to generate orders in combination with personal information. Employees can query, modify and delete books, book types, storage, orders, and their details after entering the system through their accounts. After entering the system, the administrator can query, modify and delete all the information in the current system.

1 Introduction

Now the network has more and more impact on our lives, and people can see it everywhere in our life. This huge number of software and web pages provides us with a variety of conveniences. However, in addition to fluent front-end interaction and back-end complex logic layers to achieve a variety of functions, complete back-end database support is also essential. An E-commerce platform, as an entity with huge traffic, needs to store a large amount of user data and order information every day. These data can effectively help technicians improve user experience through large data analysis, so the existence of a database is critical (Shah 2014; Monisha et al. 2014).

Z. Miao (✉)
Tianjin Sino-German University of Applied Sciences, Jincheng 048000, Shanxi, China
e-mail: 184231097@qq.com

S. Yadav et al. (eds.), *Proceedings of 2nd International Conference on Artificial Intelligence, Robotics, and Communication*, Lecture Notes in Electrical Engineering 1063, https://doi.org/10.1007/978-981-99-4554-2_22

The database is a useful tool for managing data. It is a structured data table that contains an ordered collection of data. The relationships between data tables indicate the fundamental connection between objective objects. The system administrator can successfully manage numerous information resources using the database. The database management mode enhances the security of data at the storage-level while simultaneously increasing the effectiveness of data storage. Data management is made safer and more practical by the categorized storage mode, which also provides data invocation and comparison and makes it easier to utilize queries and other operations. A product or service is an experiential good if it is consumed by the user, who can then judge its quality and usefulness. This means that in order for a consumer to determine whether to buy this good or service, she or he must rely on prior experiences because they will show if the good or service is deserving of purchase or not (Korfiatis et al. 2008; Ling 2010).

Advance management science volume 6, number1, 2017 has an article about the design and implementation of an online bookstore system. In this paper, the modules of book management, order management, user management, and sales management are involved. The system provides users with high-quality services and direct communication with the bookstore, which fully implements the original intention of system design and implementation (Han et al. 2017).

In the online database inventory for bookstore management system, written by PARISA MIRGHADERI, the problem is focused on creating a database to solve the inventory problem, which is based on the needs of the owner and the needs of the customer to implement database feathers (Mirghaderi 2009).

By consulting the related data, however, existing works are an almost single type of e-book model. In the current fast-paced society, e-books often have more advantages in some scenarios than physical books. This has led to a lot of small and medium-sized bookstores, even those with online sales systems, losing a lot of money and even being forced to go bankrupt. Therefore, this study will design a database by using SQL for an online book store to supplement the gap, while enabling the database system to effectively and safely store and manage data to meet various practical needs (Zhu 2011).

In this paper, the Online Book Sales System contains three roles of ordinary users, staff, and managers. Ordinary users have two functions: purchasing and borrowing books, returning books, and modifying personal information. First, as an online bookstore, customers can register their accounts to enter the system and supplement their personal information, so the system has a user table. When a customer chooses a book, he or she can purchase or rent an e-book or an entity book. The system then generates orders based on the information the customer fills out, so the book table and the order table are necessary, and the order type and book type tables are needed to prevent data redundancy.

Staff members have the functions of adding books, modifying book information, and modifying inventory information. The employee's information is stored in the employee table, which distinguishes the employee's work content based on the workID attribute, and the Department where the employee is located based on the department ID attribute. This requires a work table and a department table. Employees responsible for inventory can record information about each supply in the system database, including which store and which supplier it is, and modify the number of books it contains. Add book information if there are books not recorded in the database. This requires the implementation of the supplier table, storage table, and supply table.

Managers can change all the information contained in the system and have all the functions of the other two roles. The primary purpose is functional testing.

In the following content, there are five parts of the content. The second chapter explains the technology used in the system, explains it, and lists relevant examples for easy understanding. Chapter III describes the design of the system in detail and draws a system function structure diagram to provide a clearer explanation. The fourth chapter describes the database design of the system and draws out the data flow diagram, E-R diagram, and the logical structure design of the data according to the design of the system. Detailed instructions for each step are included for easy understanding.

2 The Technology Was Chosen for the Online Book Sales System

2.1 HTML Technology

Hypertext Markup Language, a markup language, is what HTML is officially known as. It contains a number of tags that can be applied to network documents to standardize their format and connect dispersed Internet resources into a logical whole. HTML text is a type of descriptive text made up of HTML commands that can display images, animations, audio, tables, links, and other media in addition to text.

Web page codes using HTML technology are implemented using templates with header content plus body content.

A tag '<html>' precedes the header content, indicating that the file is described in a hypertext markup language, that it is the beginning of the file, and that '</html>' is required at the end of the file. It is worth noting that most tags in HTML technology must appear in pairs, which is generally not recommended for tags that can be mismatched. The header content can contain information about the file, using the '<head>' tag, the '<title>' tag for the title, the '<script>' tag for the script file, and the '<style>' tag for the style file.

The body content starts with '<body>' and ends with '</body>'. The actual content displayed on the web page is contained between the two body tags.

2.2 SQL Server 2000 Technology

Relational database management system SQL Server (Structured Query Language) is a new generation of data administration and analytical software introduced by Microsoft. It is an all-encompassing, integrated, end-to-end data solution, and its purpose is to connect to and communicate with different databases (Ling and Anliang 2010).

Advanced non-procedural programming language known as Structured Query Language (SQL) enables users to work with high-level data structures. Different database systems with entirely different underlying structures can utilize the same structured query language as the interface for data input and administration because neither users' specifications nor their knowledge of how to store data are necessary. Statements in structured query languages can be nested, giving them a lot of flexibility and strength.

SQL is used because it makes it easier and faster to create and delete databases, tables, and data items. Figure 1 shows an example of creating a table.

At the same time, SQL as a language to operate the database also includes functions such as querying records, modifying records, adding fields, and so on. Figure 2 shows an example of querying a record.

Common data types in SQL are DECIMAL[(P[, S]]], FLOAT[(N)], INT, VARCHAR[(N)], DATETIME, and so on.

```
DROP TABLE IF EXISTS `department`;
CREATE TABLE `department` (
  `departmentID` int(0) NOT NULL COMMENT '部门ID',
  `departmentName` varchar(255) CHARACTER SET utf8mb3 COLLATE utf8mb3_general_ci NOT NULL COMMENT '部门名称',
  PRIMARY KEY (`departmentID`) USING BTREE
) ENGINE = InnoDB CHARACTER SET = utf8mb3 COLLATE = utf8mb3_general_ci ROW_FORMAT = Dynamic;
```

Fig. 1 An example of creating a table

```
SELECT a.S, a.Sname, cast(AVG(b.score) AS DECIMAL(18, 2)) avg_score
FROM Student a, sc b
GROUP BY a.S, a.Sname
HAVING CAST(AVG(b.score) AS DECIMAL(18, 2)) < 60
ORDER BY a.S
```

Fig. 2 An example of querying record

3 Design of Online Book Sale System

3.1 Analysis of System Functional Requirements

Online Book Sales System aims to help boost the productivity of the bookstore, delegate part of the work to the system, and reduce the workload of employees. At the same time, it improves the user experience (Han et al. 2017).

The main functions of the system are to manage, maintain and generate new data including customer information, employee information, book information, supply information, supplier information, storage information, and order information.

The System's two primary functional requirements are employee and customer needs.

3.1.1 Employee Functional Requirements

Employees have different jobs according to different work titles. The following lists the specific functional needs for employees:

Manage book information, book type, and supply information but cannot modify the ID of information. As well as only view supplier, storage, order and order type information.

Book information includes the book's ID, name, the book's author, supplier's ID, number, price, number of books available, brief, the ID of book type, storage's ID, and whether is it an e-book.

The supply information includes the ID of supply, this supply's supplier, the ID of the employee who receives, the time of supply arrival, the ID of the book, the amount of the book, the unit price of the book, the total price of books and ID of the storage stored by supply.

The supplier information includes the ID of the supplier, the name of publishing, phone number, and address. The storage information includes the ID of storage, the type of storage, and its status.

The order information includes the ID of the order, the ID of the user who placed the order, the ID of the book in the order, order generation time, ID of order type, price, return time of book, the status of books to be returned and status of the order.

There are two kinds of tables about the type, which are order type and book type. The order type information includes the ID of the order type and the name of the order type. The book type information includes the ID of the book type and the name of the book type.

3.1.2 Customer Functional Requirements

Customers can modify personal information by logging in and using functions in the system, including querying books and their information, generating orders and viewing all their order information, and modifying personal information.

In the order module, after querying to ensure that a book has an electronic version, the user can choose to rent or purchase an entity book or an e-book.

Personal information includes user name, real name, password of the account, sex, age, birthday, address, phone, code, and email.

3.2 Overview of Online Book Sale System

Online Book Sales System's main features in information management include book, book type, order, order type, employee, department, work, storage, supply, supplier, and user. The system function design chart is shown in Fig. 3.

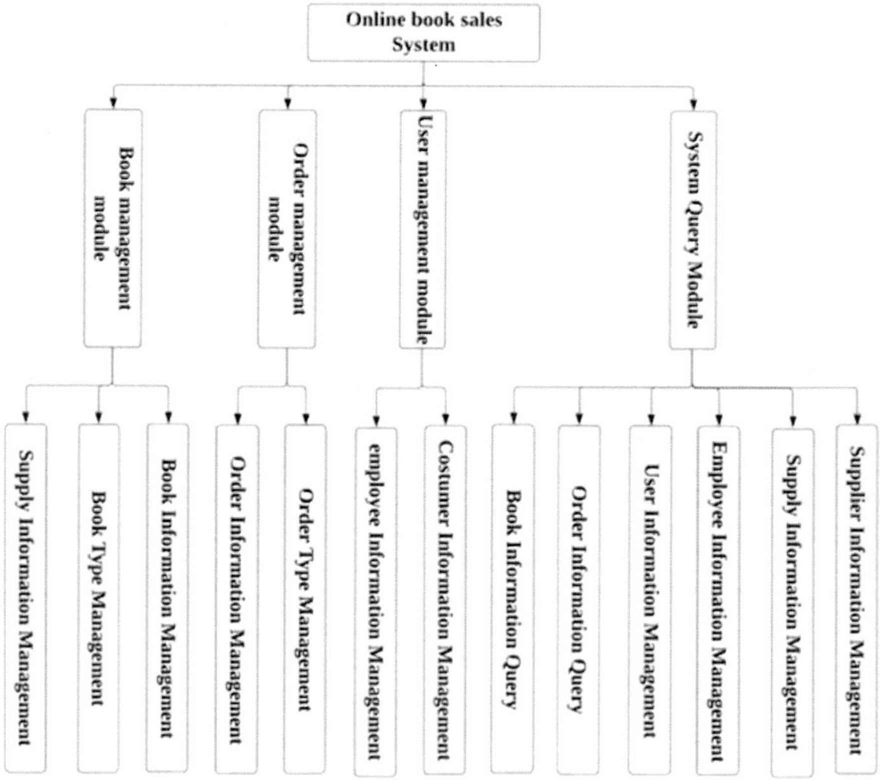

Fig. 3 System function structural

4 Database Design of Online Book Sale System

The database in the trade management system serves as the primary means of data storage and management. Consequently, it is crucial to research the system's database design and implementation (Guoming 2010).

4.1 Analysis of Database Requirement

User Requirements are reflected in providing a variety of information, saving, updating, and querying, which requires the database to fully meet the various information structures and input and output functions. Based on careful analysis and investigation of the actual situation, a data flow chart is obtained, as shown in Fig. 4.

When the user enters the system, a Customer Registration record is generated and stored in the database using the registration function (Liu et al. 2010). When a user logs in, they can use three functions: order query, book query, and purchase or rent books. When users use the function of order query, they fill in the query criteria in the search bar, and the system and the database automatically makes fuzzy queries based on the query criteria to get a record of the order information and display it on the page. When users use the book query function, they fill in keywords in the search bar, and the system will make a fuzzy query based on the keywords filled in to get related books and their information in the database. When a user purchases or rents

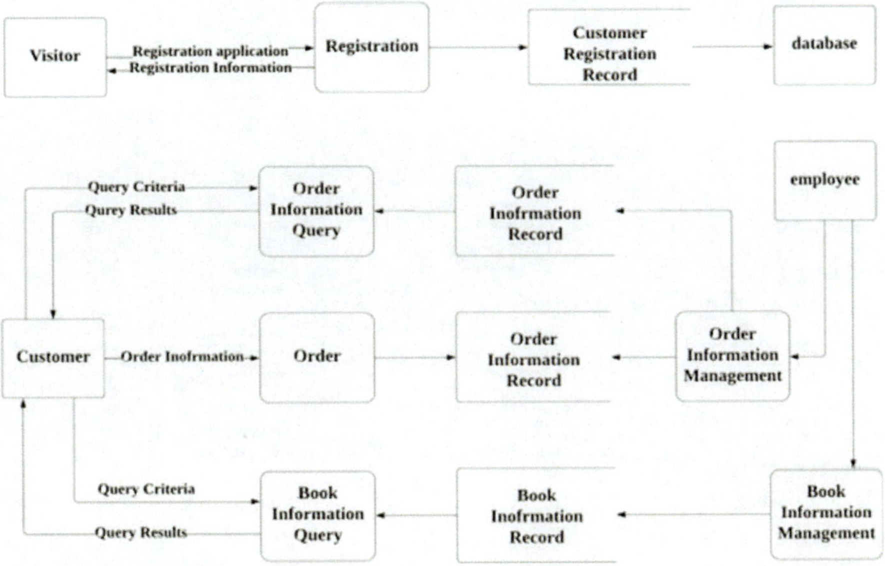

Fig. 4 Data flow diagram

a book, an order is generated based on the book and personal information the user has purchased. After confirmation, the order is saved in the database and handled by the employee.

4.2 Database Description

The design of entities and their relationships must take into account the requirements of various types of user information and serve as the design foundation for the ensuing logical structure, according to a study of database requirements. These entities interact as information flows and contain various types of particular information. There are 11 entities in the system which are book, book type, department, employee, order, order type, storage, supply, supplier, user, and work. Total entities E-R diagram shown in Fig. 5.

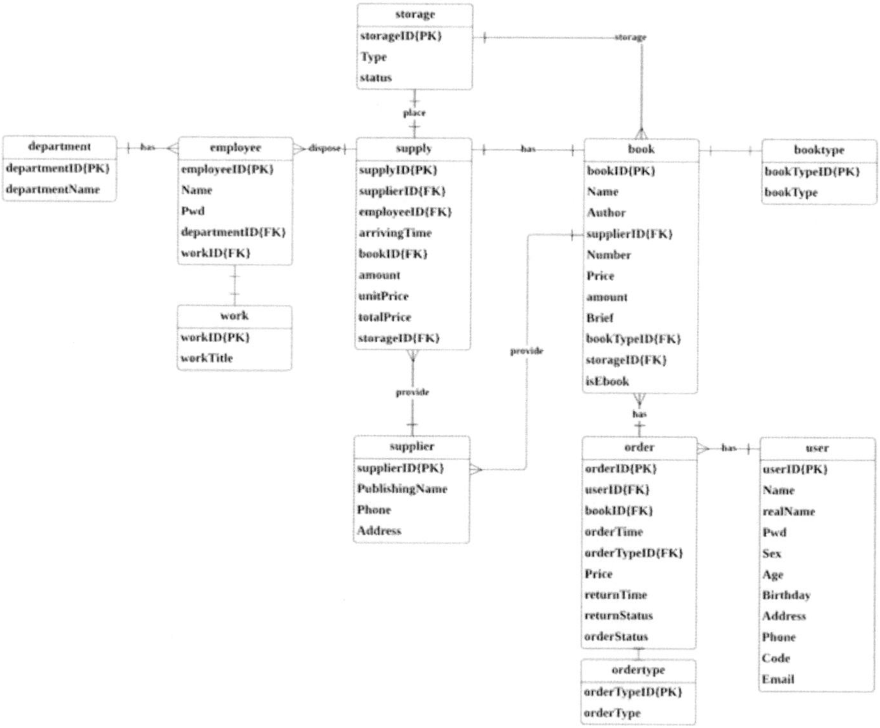

Fig. 5 E-R diagram

Table 1 Table of book basic information

Attribute	Data type	Null or NOT	Description
bookID	INT	NOT NULL	
Name	VARCHAR	NOT NULL	
Author	VARCHAR	NOT NULL	
supplierID	INT	NOT NULL	
Number	INT	NOT NULL	
Price	Decimal	NOT NULL	
amount	VARCHAR	NOT NULL	
isEbook	VARCHAR	NOT NULL	
Brief	VARCHAR	NOT NULL	
bookTypeID	INT	NOT NULL	
storageID	INT	NOT NULL	

Table 2 Table of book type information

Attribute	Data type	Null or NOT	Description
bookTypeID	INT	NOT NULL	
bookType	VARCHAR	NOT NULL	

4.3 The Logical Structure Design of the Data

The detail of book information is shown in Table 1, book type information is shown in Table 2, department information is shown in Table 3, employee basic information is shown in Table 4, order basic information is shown in Table 5, order type information is shown in Table 6, storage information shown in Table 7, supplier basic information shown in Table 8, supply basic information shown in Table 9, user-specific details shown in Table 10, work information shown in Table 11.

Table 3 Table of department information

Attribute	Data type	Null or NOT	Description
departmentID	INT	NOT NULL	
departmentName	VARCHAR	NOT NULL	

Table 4 Table of employee basic information

Attribute	Data type	Null or NOT	Description
employeeID	INT	NOT NULL	
Name	VARCHAR	NOT NULL	
Pwd	VARCHAR	NOT NULL	
departmentID	INT	NOT NULL	
workID	INT	NOT NULL	

Table 5 Table of order basic information

Attribute	Data type	Null or NOT	Description
orderID	INT	NOT NULL	
userID	INT	NOT NULL	
bookID	INT	NOT NULL	
orderTime	datetime	NOT NULL	
orderTypeID	INT	NOT NULL	
Price	decimal	NOT NULL	
returnTime	datetime	NULL	
returnStatus	VARCHAR	NULL	
orderStatus	VARCHAR	NOT NULL	

Table 6 Table of order type information

Attribute	Data type	Null or NOT	Description
orderTypeID	INT	NOT NULL	
orderType	VARCHAR	NOT NULL	

Table 7 Table of storage information

Attribute	Data type	Null or NOT	Description
storageID	INT	NOT NULL	
Type	VARCHAR	NOT NULL	
status	VARCHAR	NOT NULL	

Table 8 Table of supplier information

Attribute	Data type	Null or NOT	Description
supplierID	INT	NOT NULL	
PublishingName	VARCHAR	NOT NULL	
Phone	VARCHAR	NOT NULL	
Address	VARCHAR	NOT NULL	

Table 9 Table of supply basic information

Attribute	Data type	Null or NOT	Description
supplyID	INT	NOT NULL	
supplierID	INT	NOT NULL	
employeeID	INT	NOT NULL	
arrivingTime	Datetime	NOT NULL	
bookID	INT	NOT NULL	
amount	VARCHAR	NOT NULL	
unitPrice	Decimal	NOT NULL	
totalPrice	Decimal	NOT NULL	
storageID	INT	NOT NULL	

Table 10 Table of user basic information

Attribute	Data type	Null or NOT	Description
userID	INT	NOT NULL	
Name	VARCHAR	NOT NULL	
realName	VARCHAR	NOT NULL	
Pwd	VARCHAR	NOT NULL	
Sex	VARCHAR	NOT NULL	
Age	INT	NOT NULL	
Birthday	VARCHAR	NOT NULL	
Address	VARCHAR	NOT NULL	
Phone	VARCHAR	NOT NULL	
Code	VARCHAR	NOT NULL	
Email	VARCHAR	NOT NULL	

Table 11 Table of work information

Attribute	Data type	Null or NOT	Description
workID	INT	NOT NULL	
workTitle	VARCHAR	NOT NULL	

In the book table, the bookID serves as the primary key, and its data type is INT. This paper designed it as the store where the book is stocked is an ID given by the bookstore.

The data attributes of Name, Author, Amount, and Brief are VARCHAR, which is part of the book's attributes. The number's data attribute is INT, and it is recorded in the database as a book number that comes with the book. Price is the price of the book, and the data type is DECIMAL, because, in real life, the price of the book tends to appear after the decimal point.

IsEbook is here to determine if the book is an electronic version. The same number may appear for different bookIDs because isEbook values are different. The supplier ID, bookTypeID, and storageID are foreign keys, and the data type is INT, which represents which vendor provided them, what book type they are, and in which store they are stored.

There are two attributes in the bookType table, bookTypeID of INT type and bookType of VARCHAR type. This table is designed to prevent data redundancy in the book table, so the corresponding book type in the bookType table can be queried jointly by the foreign key bookTypeID in the book table during the query process.

There are two attributes in the Department table, the INT type departmentID, and the VARCHAR type departmentName. This table is to prevent data redundancy in the employee table. Therefore, during the query process, the corresponding department name in the department table can be queried jointly through the foreign key departmentID in the employee table.

In the employee table, the employeeID is the primary key. When an employee enters a bookstore, the table generates data about the employee through the registration function. DepartmentID and workID are foreign keys, and they represent the Department in which the employee resides and what the working title is. The three data types are INT. The data type of the Name and PWD attributes is VARCHAR.

The Order table is one of the most important tables in this system, where the orderID is the primary key, and the userID, bookID, and orderTypeID are foreign keys. Their data type is INT. The userID indicates which user-generated the order. The bookID indicates which book the order selected. Multiple data with the same orderID can appear in the table at the same time, indicating that multiple books can be selected for the same order. The purpose of the orderTypeID is to distinguish whether the order is to purchase or rent books. If the orderType found according to the orderTypeID is purchased, the returnTime and returnStatus attributes are empty. Conversely, a lease is not empty. OrderStatus represents the status of the order.

The orderTypeID is the primary key and the data type is INT. The orderType data type is VARCHAR. The table has only two pieces of data, one for purchase and one for rent.

The storageID is the primary key of the Storage table and its data type is INT. The Type attribute means the type of books put in this storage. Its data type is VARCHAR. The function of the status attribute is to confirm the status of storage, including work, repair, and full.

The data type of supplierID is INT. PublishName is the Name of the publisher, and its data type is VARCHAR. The data type of Phone and Address is VARCHAR.

The purpose of the supply table is to record information about each supply. supplyID is the primary key and its data type is INT. The foreign keys are supplierID, employeeID, bookID, storageID, which are INT types. The supplier ID can be used to find which supplier provided this provision. The employeeID refers to which employee accepts and processes the provision. The bookID refers to which book is provided. In this table, multiple pieces of data with the same supplyID are allowed, indicating that the supply provides multiple books. StorageID indicates which storage this supply will store after it is received.

The User table records all the information for registered users. The userID is the primary key for the table and its data type is INT. Name is the user's name of the account registered for the user, and realName is the user's real name. For other non-primary attributes, all are VARCHAR data types except age is of INT type.

The purpose of the Work table is to distinguish each employee's work title from the workID in the employee table. They are separated to prevent data redundancy. The data type of workID is INT, and the data type of workTitle is VARCHAR (see Table 12).

Table 12 Table of cart information

Attribute	Data type	Null or NOT	Description
userID	INT	NOT NULL	
bookID	INT	NOT NULL	
Count	INT	NOT NULL	

Whenever a user adds a book to the shopping cart, data is generated in the table to record the goods the user has selected. The userID and bookID are foreign keys, the data type is INT, the count is the number of books, and the data type is INT.

5 Implementation of the Online Book Sale System

The home page is the first page displayed when the user enters the system. Users can only browse the page without logging in, and they cannot use any function in any system. Figure 6 is the home page.

User can sign in by clicking on the sign-in button in the top menu bar. If the user has not registered an account, prompt "User not registered" after clicking the "Sign in" button on this page and close the page automatically. Figure 7 is the sign-in page.

Users can register by clicking the "sign up" button in the top menu bar to pop up the registration interface. If the registration is successful, the system will automatically close the registration page to display the home page. At this point, the user can click the "Sign in" button again to log in. Figure 8 is the sign-up page.

After a user successfully enters the system, he or she can enter keywords to search in the search bar in the upper right corner of the home page, and the search results will be displayed on the page. Figure 9 is a page display of a search result.

Users can add books to the shopping cart and modify the number of selections on the shopping cart page. On this page, the user can select some products in the shopping cart or all products to make a payment and generate an order by clicking the bottom "pay". Figure 10 is the sign-up page.

Users can click on the "My Order" button in the top menu bar to jump to the order list interface. The interface displays all orders and specific information for that user. Figure 11 is the order list page. SQL statement to get all orders is as the Fig. 12. SQL statement to get query orders is as the Fig. 13.

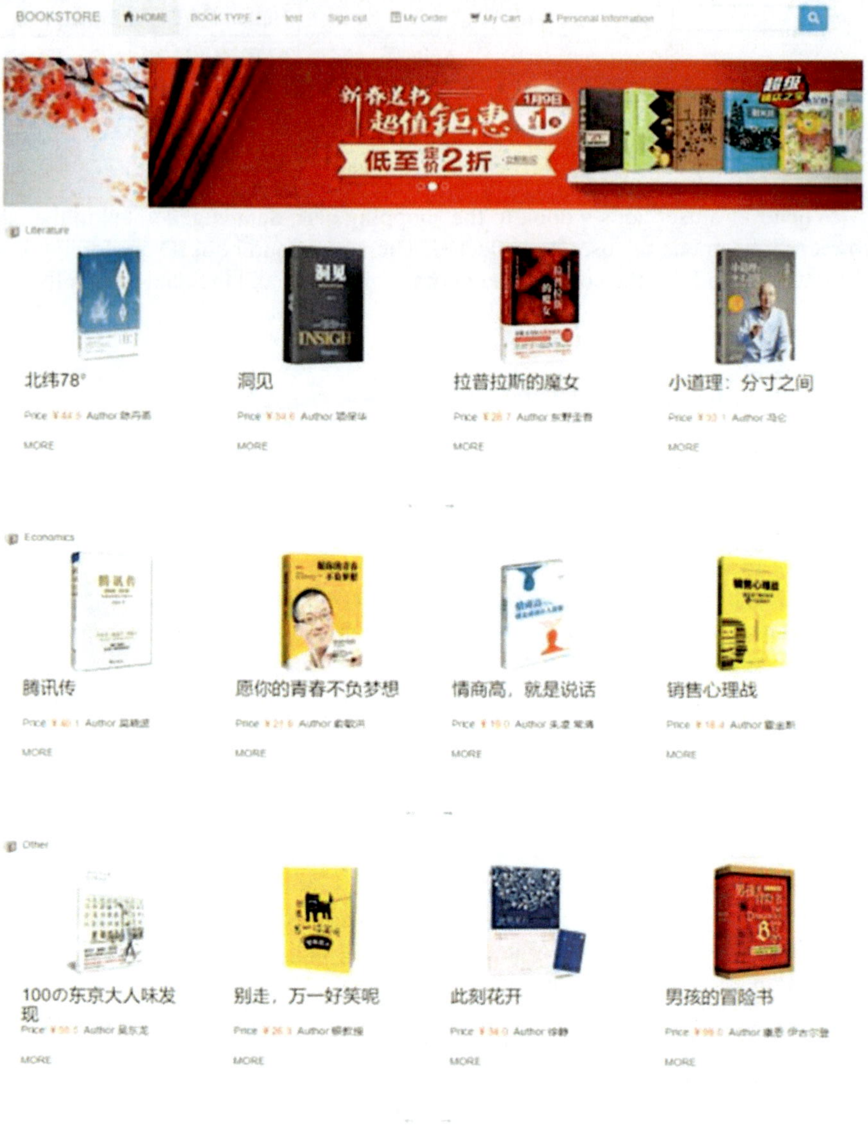

Fig. 6 Home page

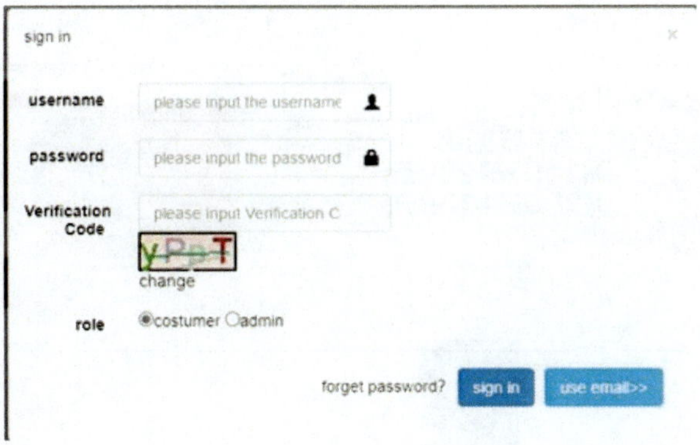

Fig. 7 Sign in page

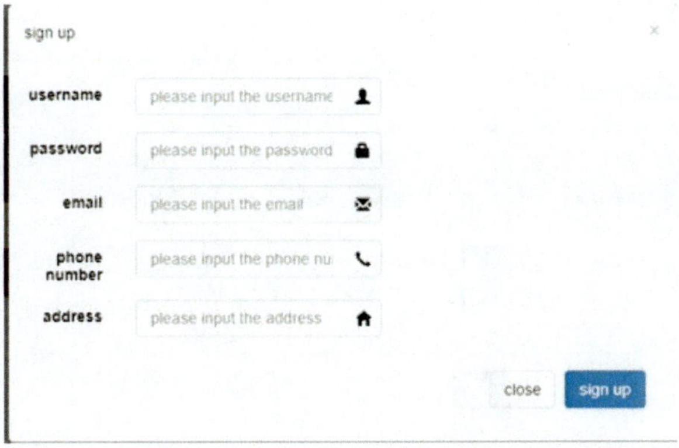

Fig. 8 Sign up page

Fig. 9 Search result page

Fig. 10 Cart page

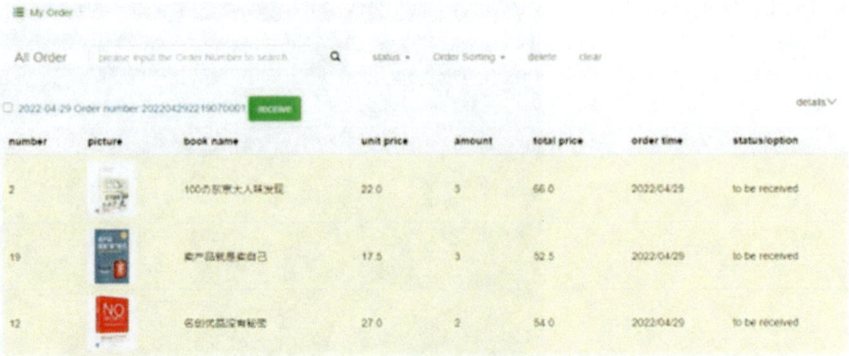

Fig. 11 Order list page

```
<select id="findOrderAndOrderDetailListByUser"
        parameterType="OrderQueryVo"
        resultMap="orderAndOrderDetailResultMap">
    SELECT bo.*, bu.username, bu.phone, ba.province,
    ba.city, ba.area, ba.detail_address, ba.tel, ba.receiver,
    boi.id order_item_id,boi.book_id,boi.order_id,
    boi.count,boi.state,boi.price,boi.single_price,
    bb.id book_id,bb.img_url,bb.name,bb.new_price
    FROM
    bs_order bo,bs_user bu,bs_address ba,bs_order_item boi,bs_book bb
    <where>
        <include refid="query_order_where"></include>
        <if test="begin != null and end != null">
            limit #{begin},#{end}
        </if>
    </where>
</select>
```

Fig. 12 The SQL about getting all orders

Fig. 13 The SQL about getting query orders

```
<sql id="query_order_where">
    bo.user_id = bu.id
    AND bo.address_id = ba.id
    AND bo.id = boi.order_id
    AND boi.book_id = bb.id
    <if test="userId != null">
        and bo.user_id = #{userId}
    </if>
    <if test="orderNum != null and orderNum != ''">
        and bo.order_num = #{orderNum}
    </if>
    <if test="tradeStatus != null and tradeStatus != ''">
        and bo.order_status = #{tradeStatus}
    </if>
    <if test="startTime != null">
        and bo.create_date > #{startTime}
    </if>
    <if test="endTime != null">
        and bo.create_date < #{endTime}
    </if>
    <if test="orderBy != null and orderBy != ''">
        order by bo.create_date ${orderBy}
    </if>
</sql>
```

6 Conclusion

In order to supplement the missing e-book module and rental module in the ordinary bookstore system, this paper implements an online book sales system. If the system is put into use, it will produce a large amount of data every day, which needs to be stored and managed by the database, then operated by the back-end logic layer to provide users with high-quality services. Therefore, the design and implementation of the database is necessary. The design of the database can meet the requirements of system function. The system realizes the basic functions of user management, order management, commodity management, and designs different roles with different functions according to the actual situation. For customers, after registering with the login system, they can read books and their details, and if they are interested, they can join the shopping cart. Users can purchase or rent books directly in the shopping cart or directly on the book details page, thereby generating orders and storing them in the database. For bookstore employees, the system will show different functions according to different work titles. In this system, employees can manage supply, manually import supply specific information into the database and perform various operations in the system. Books can be managed and staff can modify, add or delete book information as needed. For an incomplete order, the system will automatically follow up on the information in the order and modify the order information in time. In addition to the customer and employee roles, there is also a system administrator role. This role is typically used for system function testing and has privileges to use all functions in the system. This series of functions enables effective data management and greatly improves employee productivity, while enabling customers to use the system more easily. The system also extends the capabilities of e-books and leasing, which have great advantages for the current environment. E-books now have a larger user base, making them more convenient and affordable. And the leasing function can provide a more affordable solution, and the book can still be used with less money. The system has good scalability and can be redeveloped according to new features in a specific application environment. However, the design and implementation of this system involves many aspects of theory, method and technology. There are many new problems to be solved in this system, which need to be accumulated and perfected in practical application, and further research and development are needed.

References

Guoming Y (2010) The general principles of relational database design. In: Seventh China institute of communications annual conference proceedings

Han M, Tan D, Zhang Y, Zhang Y (2017) The design and implementation of online bookstore. Int J Adv Manag Sci 6(1):73–75

Korfiatis N, Rodríguez D, Sicilia MA (2008) The impact of readability on the usefulness of online product reviews: a case study on an online bookstore. Lecture notes in computer science book series. LNAI, vol 5288, pp 423–432

Ling W (2010) The CMM-based software development platform of adapted to the information system. Inf Syst Eng

Ling X, Anliang Z (2010) Relations of information technology and human resources management. Mark Mod

Liu Z, Wang H, Zan H (10 Dec 2010) Design and implementation of student information management system. In: IEEE 2010 international symposium on intelligence information processing and trusted computing

Mirghaderi P (2009) Online database inventory for bookstore management system. Unpublished

Monisha S, Sangeetha K,Annapoorna S (May 2014) Online book sales in e-commerce with multilingual support and mobile alert. Int J Eng Res Sci Technol 3(2)

Shah P (2014) Online bookstore. Unpublished

Zhu L (5 May 2011) Research and application of SQLserver in the trade management system. In: IEEE 2011 3rd international conference on computer research and development, China

Review of Tobacco Planting Area Estimation Based on Machine Learning and Multi-source Remote Sensing Data

Ronggang Gao and Fenghua Huang

Abstract Tobacco is an important cash crop in China. The income of tobacco industry is an important part of national economic income. Planting tobacco has become an important way to solve the employment problem of rural population and increase farmers' income in tobacco growing areas. The tobacco planting area is large and the scope is wide. It is one of the important research fields of tobacco planting and production to master the growth status of tobacco leaf in real time and in a large area. At present, the supervision of tobacco planting mainly depends on artificial ground survey, which is time-consuming, laborious and difficult to be carried out in many places at the same time. Remote sensing has the characteristics of large area synchronous observation, fast data acquisition, short cycle, etc. It has been widely used in geological exploration, environmental monitoring, agriculture and animal husbandry and other fields. In recent years, more and more researchers have applied remote sensing to tobacco identification and planting area estimation. This paper comprehensively applies multi-source remote sensing data such as satellite images and UAV images to analyze and summarize the application status of remote sensing in tobacco identification and planting area estimation, and prospects for further research.

R. Gao
Digital China Research Institute (Fujian), Key Laboratory of Spatial Data Mining&Information Sharing of Ministry of Education (Fuzhou University), Fuzhou 350108, China

F. Huang (✉)
Fujian Key Laboratory of Spatial Information Perception and Intelligent Processing (Yango University), Fuzhou 350015, China
e-mail: fenghuait@sina.com

Fujian University Engineering Research Center of Spatial Data Mining and Application (Yango University), Fuzhou 350015, China

S. Yadav et al. (eds.), *Proceedings of 2nd International Conference on Artificial Intelligence, Robotics, and Communication*, Lecture Notes in Electrical Engineering 1063, https://doi.org/10.1007/978-981-99-4554-2_23

1 Acquisition of Multi-source Crop Remote Sensing Information

China is the country with the largest tobacco planting area in the world. Tobacco planting is mainly distributed in the southwest of China, with the planting area reaching 66.1% (Wang et al. 2021). Tobacco is usually grown once a year, usually in spring (Zhao 2017). In order to ensure economic benefits, most tobacco plants are planted in rotation, mainly including paddy and dry rotation and dry land rotation. Due to the small per capita planting area in China, tobacco planting land is scattered, which is difficult to intensive management (Wu 2018). The tobacco production area has moved from the plain to the backward mountain area (Li et al. 2021). It takes remote sensing images as information sources and provides decision-making support for agricultural production through the processing and analysis of remote sensing image information (Chen et al. 2016). In particular, the advantages of high resolution images are more prominent, which makes it possible to finely classify and monitor ground objects (Shen et al. 2016).

1.1 Satellite Remote Sensing Data Acquisition

Since the 1970s, foreign satellite remote sensing data, such as Landsat TM and SPOT, have been generally used for crop area estimation (Kuenzer and Knauer 2013; Duveiller and Defourny 2010). Remote sensing data from low resolution NOAA-AVHRR, MODIS (Peña-Arancibia et al. 2014; Massey et al. 2017) and FY series satellites for large-scale agricultural remote sensing monitoring (Doraiswamy et al. 2003). Generally, Landsat TM and SPOT data are used for data fusion to improve temporal and spatial resolution before tobacco extraction (Wu et al. 2008; Xin et al. 2006). MODIS data is generally used for multitemporal analysis of tobacco crop spectrum (Wu and Cui 2007). In recent years, with the rapid development of remote sensing satellites and observation technology, more and more medium and high-resolution remote sensing data sources have been provided, and the scale of data open to the public is also growing. Table 1 is a brief introduction to various satellite data sources that can be applied to tobacco crop extraction.

1.2 UAV Aerial Remote Sensing Data Acquisition

With the rapid development of remote sensing, global positioning system and other technologies, UAV aerial remote sensing has made great progress, providing technical support for the further development of precision agriculture. It can improve the ground crop monitoring system, and play a greater role in the application of small-scale agricultural remote sensing. In the field of monitoring economic crops (such as

Table 1 Introduction of satellite parameters

Data source	Band range (μm)	Minimum spatial resolution (m)	Product spatial resolution range (m)
Landsat TM/ETM+	0.45–12.5	30	30
MODIS	0.4–14.4	250	250–5600
SPOT	0.455–0.89	1100	1100
Sentinel-2	0.423–2.270	10	10–60
ZY-3	0.45–0.89	2.1	2.1–5.8
CBERS	0.5–0.89	2.36	2.36–10

tobacco) with complex planting structure, UAV remote sensing can provide clear and accurate image information, providing strong data support for tobacco identification and area estimation (Chen et al. 2018; Xiang and Tian 2011; Candiago et al. 2015).

2 Tobacco Identification and Planting Area Calculation

At present, the method of crop area extraction is relatively mature, and has also been greatly applied in tobacco crop extraction, which can be summarized as supervised classification and unsupervised classification. When it comes to specific classification problems, supervised classification is usually used, which can be subdivided into three main classification methods, including pixel-based, object-oriented and neural network-based classification methods. For medium and low remote sensing images with low spatial resolution, image fusion is usually used to improve the resolution (Peng et al. 2009).

2.1 Pixel-Based Classification Method

The traditional crop remote sensing classification method mainly takes pixel as the research object, fully excavates spectral information, and uses band information and normalized vegetation index (NDVI), difference vegetation index (DVI), ratio vegetation index (RVI) and other indexes to extract ground objects.

(1) Decision tree algorithm

Decision tree algorithm is a method of approximating the value of discrete function. The decision tree is built by analyzing the crop spectral curve and vegetation index features to extract classification rules. Zhang et al. (2020) selected Chaling County, Hunan Province, as the research area to study the effect of Sentinel-2A data on remote sensing monitoring of flue-cured tobacco planting area. The algorithm based on decision tree can organically combine remote sensing, crop phenology and expert knowledge, avoid logical classification errors, and effectively improve the classification accuracy (see Fig. 1).

(2) Random forest algorithm

It is an algorithm with integrated learning idea. It learns multiple decision trees through training. When the prediction is needed, the results of multiple decision trees are integrated in a certain form as the final result output. It effectively improves the universality and robustness of the algorithm, and has better prediction performance than a single decision tree. The algorithm process is shown in Fig. 2. This algorithm is widely used in crop extraction and classification (Peng et al. 2021; Wu et al. 2020; Wang 2019). The specific advantages of this algorithm are shown in Table 2.

(3) Other classification algorithms

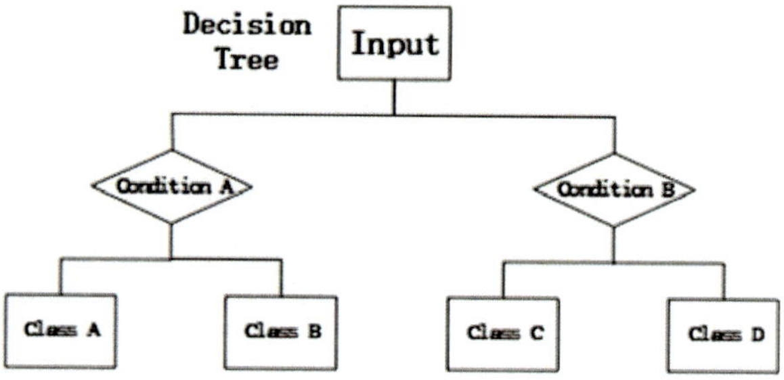

Fig. 1 Decision tree

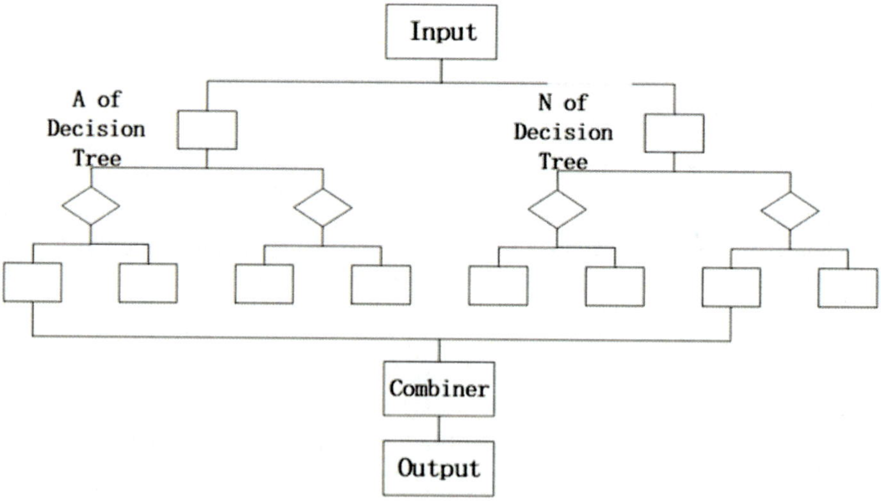

Fig. 2 Random forest algorithm

Table 2 Comparison of advantages and disadvantages of random forest classification

Advantage	Disadvantage
Handle high-dimensional features without feature selection	The model is not very explainable
Loss of dataset features still maintains model stability	When the dataset is small or has few features, classification performance is not guaranteed
The model has strong generalization ability	Training is slow compared to a single decision tree
Ability to solve classification and regression problems	It is easy to overfit on some noisy classification or regression problems

Xin et al. (2006) used pixel based maximum likelihood classification to extract tobacco from TM and SPOT fusion images. Wu et al. (2008) extracted NDVI and gray level symbiotic texture matrix from TM and SPOT fusion images, extracted crops layer by layer using the threshold method. Wang et al. (2014) and others also analyzed the spectral characteristics and vegetation index differences between flue-cured tobacco and other crops, selected DVI as the classification index, and extracted flue-cured tobacco through the threshold method.

The above methods are all based on spectral features and combination index, only considering a small number of temporal images, not considering the temporal characteristics of tobacco and other crops, and the selection of classification features is too few, which makes it difficult to ensure the robustness of the model.

Pixel-based classification methods usually produce salt and pepper phenomenon, especially when ground objects are fragmented and distributed. Aiming at the problem that pixel-based classification methods are prone to salt and pepper, researchers have proposed an object-oriented method.

2.2 *Object-Oriented Classification Method*

The operation scale unit of the object-oriented method is not based on a single pixel, but is based on the image object (Shi et al. 2012). Image object can use random forest, decision tree and other algorithms for direct classification, and can also combine expert knowledge and auxiliary data for better classification. Based on different remote sensing data sources, object-oriented methods of tobacco extraction have achieved good results. Li (2013a) used the fusion image of CBERS to perform multi-scale segmentation and then feature selection and analysis, and obtained the tobacco distribution map through object-oriented decision tree classification. Mei et al. (2014) and others successfully extracted the tobacco of Linyi Ten Thousand Mu Demonstration Park from the UAV orthophoto with the object-oriented method of eCognition. Based on multi-temporal environmental satellite data and smoothly constructed NDVI time series curve data, Zhang et al. (2015) adopts decision tree classification at the object scale. Based on the panchromatic band and multispectral band of the single-phase ZY-3 satellite, Liu (2016) extracted tobacco planting area and estimated yield in Yishui County, Linyi, Shandong Province. The object-oriented classification method is used to segment the image and extract tobacco crop. Compared with the decision tree classification and unsupervised classification results, the object-oriented method achieves better results. Yufei et al. (2022) used Sentinel-2 multispectral data as the data source, analyzed the spectral characteristics, woodland, water body and other ground objects in March 2020, and applied the object-oriented nearest neighbor algorithm to accurately extract the flue-cured tobacco planting area information.

2.3 *Classification Method Based on Neural Network*

Neural network is a widely parallel and interconnected network composed of adaptive simple units. It implements end-to-end tasks. Only input variables are needed to get the output. In the past ten years, deep learning algorithms have been greatly developed. It is widely used in scene segmentation, speech recognition, object classification, automatic driving and other fields. Convolutional neural network (CNN) is very effective in object classification, and full convolutional neural network (FCN) has also made great progress in scene segmentation. Because the semantic segmentation in computer vision is similar to the classification of remote sensing images, researchers began to introduce FCN to learn the local and global features of remote sensing images. In the field of tobacco crop recognition, Fan et al. (2018) proposed for the first time a UAV tobacco detection method based on depth learning, combined with morphological segmentation and watershed algorithm, to achieve automatic tobacco counting. Lei et al. (2021) used the classic FCN algorithm (U-net network) to achieve tobacco extraction in the study area. Fu and Liang (2022) proposed an accurate extraction method of tobacco planting area from UAV remote sensing images based on DeeplabV3+ depth semantic segmentation model. U-net, DeeplabV3+ and other networks integrate high-level semantic features and low-level fine features through multi branch structure, effectively improving the edge thinning effect and improving the classification accuracy. FCN provides an effective solution for accurate extraction and area monitoring of tobacco crops (Tian 2022). Figure 3 shows the network architecture of the typical FCN algorithm—Deeplab V3+. Tobacco crops are mostly planted in mountainous areas, and the planting area is relatively small compared with forest land, grassland and other vegetation. In image recognition, high-level features should be retained as well as low-level features to improve the accuracy of tobacco extraction. Therefore, we can further study the performance of related FCN algorithms in tobacco extraction based on different data sources.

Typical classification algorithms based on pixel, object-oriented and depth learning are summarized in Table 3.

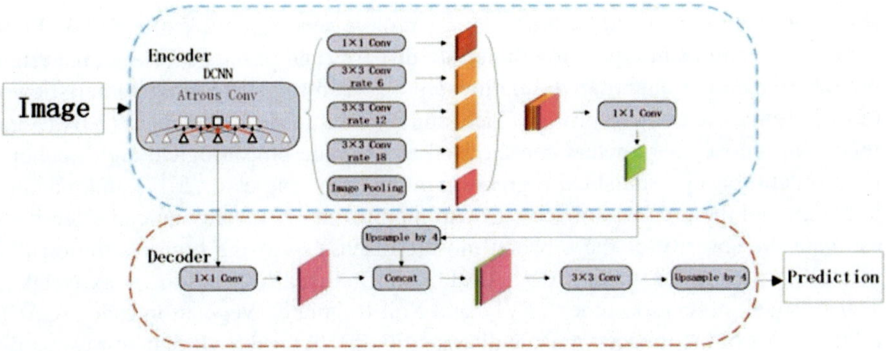

Fig. 3 Deeplab V3+

Table 3 Summary of classification methods of related documents

Literature	Classification method	Data source	Summary
Xin et al. (2006)	Cell-based	TM, SPOT, SAR	By multi-source data fusion, the image resolution is improved, but it is easy to produce pepper and salt phenomenon
Wu et al. (2008), Zhang et al. (2020)		TM, SPOT, Sentinel-2A	
Wang et al. (2014)		HJ-1	
Li (2013a), Liu (2016)	Object-oriented	CBERS, ZY-3	It makes full use of the spatial, structural and texture information, avoids the phenomenon of salt and pepper
Mei et al. (2014), Yufei et al. (2022)		UAV, Sentinel-2	
Fan et al. (2018)	Deep learning	UAV	An end-to-end classification algorithm is proposed, which promotes the automatic extraction process of tobacco crops
Lei et al. (2021)		GF-2	
Fu and Liang (2022)		UAV	

3 Analysis of Tobacco Growth

The quality and growth of tobacco planting are affected by many factors, including soil, light and temperature (Chen et al. 2021). Tobacco has typical spectral characteristics of green vegetation (Shen et al. 2016). The reflectance spectrum of tobacco leaves or canopy obtained by multi-source remote sensing data is affected by light, variety, moisture, soil type, growth period, diseases and pests, etc., showing certain differences, so the quantitative relationship between the spectral characteristics of tobacco reflectance and its physiological and biochemical parameters Li (2013b) and related agronomy parameters can be used to estimate physiological and biochemical content through statistical regression methods (Liang et al. 2017) and machine learning, and then realize tobacco growth monitoring. Xu Dongyun can effectively estimate the severity of the tobacco mosaic disease by using high-resolution ZY-3 satellite images and taking RVI, DVI, Renormalized Vegetation Index (RDVI), Transformed Vegetation Index (TVI), and Soil Regulated Vegetation Index (SAVI) as indicators of the tobacco mosaic disease. Fu (2015) used SAR data to reverse the leaf area index of tobacco, combined with the growth status parameters such as the leaf area index of tobacco, the suitability of growth environment parameters, etc., to establish a comprehensive yield estimation model for tobacco in Guizhou plateau mountainous areas.

4 Summary and Prospect of Tobacco Remote Sensing Research

Multi-source and multi-platform remote sensing technology can effectively obtain tobacco information, providing strong technical support for tobacco growth analysis and area extraction. This paper focuses on the application of remote sensing data sources to tobacco crop agriculture. Based on the data format and the development trend of current machine learning methods, this paper summarizes the classification methods based on pixel, object-oriented and deep learning. These three methods can effectively extract tobacco crop area, but some methods have some limitations, and need to choose appropriate classification methods according to the research area and data sources. Therefore, there are three prospects:

(1) At this stage, tobacco physiological and biochemical indexes and related parameters are mainly retrieved through spectra and correlation indexes, which are quantitative inversion based on statistical models. Due to regional restrictions and different remote sensing sources, their universality is poor. In order to solve this problem, it is necessary to describe the internal logical relationship between tobacco parameters, spectrum and index, and introduce physical parameters to establish a physical model or semi-empirical model. At present, the research on the remote sensing mechanism is very complex, involving too many physical parameters, easy to over fit, and consuming more computing resources, which is very challenging.
(2) Tobacco crops are usually planted in a decentralized manner, and the area of a single plot is small. In order to improve the growth and yield of tobacco, rotation or intercropping is usually selected, which requires higher resolution

of remote sensing data, higher spatial resolution and time resolution. Landsat-TM, Sentinel can provide certain support. Sentinel can obtain high resolution remote sensing images for free. In the future, tobacco extraction research can try to use more Sentinel satellites to give full play to the role of Sentinel satellites in crop extraction.

(3) The traditional pixel-based classification method does not consider the relationship between adjacent pixels of actual objects, which violates the law of "The more similar geographic configurations of two points (areas), the more similar the values (processes) of the target variable at these two points (areas)" in geography, especially in the extraction of complex objects such as tobacco. The object-oriented classification method solves this problem well. However, these classification methods need to focus on the analysis of spectral characteristics, various vegetation indexes and time series characteristics, which increase the workload of classification. Deep learning is end-to-end learning and does not need to pay attention to internal features. At present, crop extraction based on deep learning has been studied maturely, but it mainly focuses on the classification and recognition of bulk crops, and the research area has neat plots and simple planting structure. There are few studies on the extraction of crops with complex planting structure, such as tobacco. The next step should focus on the application of deep learning in tobacco extraction.

Acknowledgements This work was funded by the National Natural Science Foundation of China (NSFC, 41501451), Natural Science Foundation of Fujian Province in China (No.2019J01088). The author would like to thank Fenghua Huang in the Spatial Information Research Center (SIRC) of Fujian Province (China) for him assistance, suggestions, and discussions.

References

Candiago S et al (2015) Evaluating multispectral images and vegetation indices for precision farming applications from UAV images. Remote Sens 7(4):4026–4047

Chen Z et al (2016) Progress and prospect of agricultural remote sensing research and application. J Remote Sens 20(05):748–767

Chen J et al (2018) Application status of tobacco field monitoring technology based on UAV remote sensing. Mod Agric Sci Technol 04:211–212

Chen C et al (2021) Analysis on the application of key technologies for high-quality and efficient cultivation of tobacco. Seed Sci Technol 39(21)45–46

Doraiswamy, Paul C., et al. "Crop yield assessment from remote sensing." Photogrammetric engineering & remote sensing 69.6 (2003): 665 674.

Duveiller G, Defourny P (2010) A conceptual framework to define the spatial resolution requirements for agricultural monitoring using remote sensing. Remote Sens Environ 114(11):2637–2650

Fan Z et al (2018) Automatic tobacco plant detection in UAV images via deep neural networks. IEEE J Sel Top Appl Earth Obs Remote Sens 11(3):876–887

Fu B, Liang H (2022) Extraction of tobacco planting area from UAV image based on depth semantic segmentation. Commun Technol 55(02):181186

Fu Y (2015) Research on the application of high resolution spaceborne SAR in the estimation of tobacco yield in plateau mountainous areas. Guizhou Normal University, Matthesis

Kuenzer C, Knauer K (2013) Remote sensing of rice crop areas. Int J Remote Sens 34(6):2101–2139

Lei Z et al (2021) Extraction of tobacco planting information based on U-net neural network. Agric Technol 41(22):44–47

Li T (2013a) Research on extracting tobacco planting area based on object-oriented classification method. Sichuan Agricultural University

Li J (2013b) Effects of different nitrogen levels on hyperspectral characteristics and physiological and biochemical indexes of flue-cured tobacco. Henan Agricultural University, Matthesis

Li L et al (2021) Current situation and development trend of tobacco planting machinery in Guangdong. Mod Agric Equip 42(06):80–84

Liang B et al (2017) Study on the relationship between tobacco leaf quality and soil trace element content in Honghe tobacco planting area. Southwest Agric J 30(04):824–829

Liu M (2016) Study on the estimation model of intercropping tobacco yield in complex mountain areas based on the remote sensing image of resource 3. Shandong Agricultural University

Massey R et al (2017) MODIS phenology-derived, multi-year distribution of conterminous US crop types. Remote Sens Environ 198:490–503

Mei D et al (2014) Object oriented UAV remote sensing image tobacco planting area extraction and monitoring. Surv Mapp Sci 39(09):87–90

Peña-Arancibia JL et al (2014) Dynamic identification of summer cropping irrigated areas in a large basin experiencing extreme climatic variability. Remote Sens Environ 154:139–152

Peng G et al (2009) Remote sensing monitoring of tobacco field based on phenological characteristics and time series image—a case study of Chengjiang county, Yunnan province, China. Chin Geogr Sci 19(2):186–193

Peng Z et al (2021) Research on crop classification based on time series Sentinel-2 image. Mapp Spat Geogr Inf 44(12):81–84

Shen X et al (2016) Overview of the application of hyperspectral remote sensing technology in tobacco. Jiangxi J Agric 28(07):78–82

Shi XL, Li Y, Deng RX (2012) Object-oriented information extraction of farmland shelterbelts from remote sensing image. Key Eng Mater 500. Trans Tech Publications Ltd.

Tian T et al (2022) Research on fine classification of crops in areas with complex planting structure based on deep learning model. Agricultural Resources and Zoning in China

Wang R (2019) Research on crop planting structure extraction method based on the synthesis of spectral and texture features. Lanzhou Jiaotong University, Mathesis

Wang Z et al (2014) Identification method of flue-cured tobacco based on spectral features in HJ-1 image. Tob Sci Technol (01):7276

Wang Z et al (2021) Research status and suggestions on tobacco mechanization in mountain and hilly areas. Agric Eng 11(12):20–25

Wu Z (2018) Tobacco planting system and its technical guarantee measures. Agric Technol 38(14):11–12

Wu M, Cui W (2007) Monitoring the growth of large-area tobacco with MODIS data. In: MIPPR 2007: remote sensing and GIS data processing and applications; and innovative multispectral technology and applications, vol 6790. SPIE

Wu M et al (2008) Research on remote sensing monitoring and information extraction methods of tobacco planting in complex mountain areas. Remote Sens Technol Appl (03):305–309

Wu L et al (2020) Crop hyperspectral remote sensing recognition based on random forest method. J Zhejiang Agric For Univ 37(01):136142

Xiang H, Tian L (2011) Development of a low-cost agricultural remote sensing system based on an autonomous unmanned aerial vehicle (UAV). Biosys Eng 108(2):174–190

Xin M et al (2006) Monitoring of tobacco planted acreage based on multiple remote sensing sources. 2006 IEEE international symposium on geoscience and remote sensing. IEEE

Yufei X et al (2022) Accurate extraction of tobacco planting information based on Sentinel-2 remote sensing image. China Tob Sci 43(01):96–106

Zhang H et al (2015) Research on crop classification based on multi temporal environment star NDVI time series. Remote Sens Technol Appl 30(02):304–311

Zhang Y et al (2020) Extraction and analysis of county flue-cured tobacco planting area based on Sentinel-2A data. Tob Sci Technol 53(11):15–22

Zhao L (2017) Application of tobacco planting technology and field management. Agric Technol Serv 34(24):53

An Innovative Information System for Health and Nutrition Guidance

Yufei Li and Yuqing Li

Abstract The restrictions of the epidemic have had a significant impact on people's habits, most notably changes in eating habits, which are very harmful to human health. Issues such as the epidemic lockdown have reduced the opportunity for people to eat out, and this series of changes has led to a decrease in concern for the intake of nutritious and balanced foods, which has had a negative impact. The human immune system is closely related to the intake of healthy foods, and a balanced diet will ensure a strong immune system, and as a restaurant industry there is an opportunity to help people get healthier meal combining recommendations through an efficient and versatile restaurant management system. This article develops a restaurant management system that adds consideration of customer eating habits and recommendations for healthy meal combinations to the basic restaurant management functions, including the creation of a health profile function, nutritional advice and recommendations for dishes and combinations for personalized orders.

1 Introduction

Good nutrition and eating habits are vital to human's health. Especially during the pandemic, people's eating habits and lifestyle are closely related, and the immune system of the human body is also closely related to the intake of healthy food. The lockdown and a series of problems brought by COVID-19 have greatly reduced people's willingness to go out and the opportunity to dine in restaurants. The reduction of human activities has increased the risk of cardiovascular disease. According

Yufei Li and Yuqing Li are contributed equally.

Y. Li
Eller College of Management, University of Arizona, Tucson, USA
e-mail: yufeili@arizona.edu

Y. Li (✉)
School of Public Health, University of Washington, Seattle, USA
e-mail: yli37@uw.edu

S. Yadav et al. (eds.), *Proceedings of 2nd International Conference on Artificial Intelligence, Robotics, and Communication*, Lecture Notes in Electrical Engineering 1063, https://doi.org/10.1007/978-981-99-4554-2_24

to Wang et al. (2021), the negative emotions brought about by the pandemic will lead to an increase in the consumption of foods with high sugar and energy, and people tend to buy foods with a long shelf life and usually high salt and fat. In addition, according to Aman and Masood (2020), the behavior of staying at home can have a significant impact on a person's health, including changes in eating patterns, sleep habits and physical activity. A balanced diet will ensure a strong immune system that can help ward off viral attacks. The restaurant industry should respond to this phenomenon and change people's eating habits by adding dietitian functions to restaurant management systems. The existing restaurant management system functions have been basically perfected, and all the required functions have been realized for the restaurant managers.

The existing catering management system has been functionally perfected, and a standardized management system has been built for restaurant managers and restaurant administrators, which has significantly improved the restaurant's operational efficiency (Kurniawan et al. 2019). Although it is complete in some ways, it still does not take customers' eating habits into consideration. Considering the beneficial effects of healthy eating habits and nutritionally balanced meals on the body, this restaurant management system adds consideration of customers' eating habits and recommendation of healthy meal matching to the basic restaurant management functions.

Undernutrition and nutritional imbalance are important causes of many diseases (World Health Organization 2021). The body needs nutrition all the time, and the growth and development of people, health level, labor ability and length of life are closely related to nutrition (Lifshitz 2021). Nowadays, many people are aware of the importance of nutrition but still do not know enough about how to understand and arrange their diet according to their personal situation. It would be an effective tool for people to manage their health if there was a restaurant system linked to nutrition and health to help people understand and make better food choices.

Now is the era of data. In order to collect and better manage data and provide services to people, many departments will establish databases and information systems. The existing public health authorities are starting to figure out how to build and optimize information systems to make better use of all that information (Revere and Turner 2007). This paper aims to design a system based on the existing restaurant system for an innovative combination of restaurant and nutrition advice services. The system has the control and management of the supply chain, storage, employee, order and membership that a popular restaurant needs. On this basis, this paper first added the function of new health files for members. A registered member can set up multiple health profiles to receive dietary advice from dietitians. This health file can be the registered member's own file or his family or friends. After completing the input of the archival information, the system will provide nutritional advice and recommended dishes and combinations in the restaurant according to the information. Customers can make personalized orders based on the recommended dishes.

New menu categories are also added to help customers choose. The menu can be differentiated according to the type of dish (such as main course and dessert) (this is the same as a normal restaurant system). The menu can also be differentiated according to the type of diet (such as the low-fat diet) and the customer's taste preference (such as spicy or sweet).

2 Design of Health System

2.1 System Function Structure

The function of this system is to support the entire operation of the restaurant while providing nutritional and health advice to customers (shown in Fig. 1). When operating a restaurant, restaurant managers need to manage and control supply chain information, employee file information, registered customer information, menu information and order information. These are the basic functions of the restaurant system. In addition to this, this article designed a new system to provide customers with dietary nutrition and health advice. According to the input of the customer's personal body information, corresponding suggestions will be added.

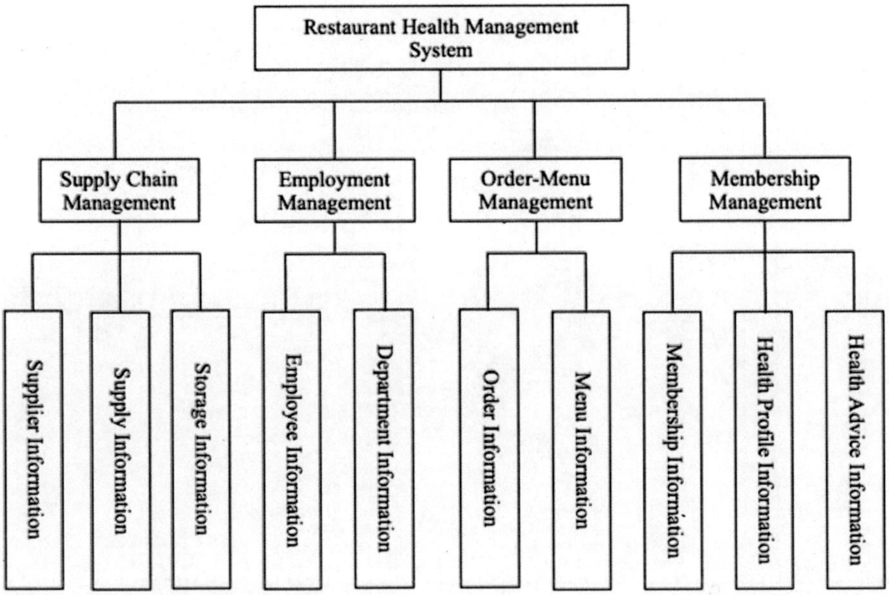

Fig. 1 System function structure

2.2 Database Description

When designing a software system, the database is an integral part of the software system. Before creating a database, designers usually create an ER diagram which is a kind of structure diagram that can ensure high quality database design for database creation, management, and maintenance. ERD visualizes the main entities in the system, such as business objects, and the relationships between these entities. The designed database includes 16 entities (shown in Fig. 2).

(1) *Supply Chain Information Management*

The supply chain system (shown in Fig. 3) behind the restaurant will meet the restaurant's needs for necessary elements such as food materials, restaurant supplies, etc. This part of the ER diagram about supply chain information management includes three necessary entities: supplier (vendor), supply (supply detail), storage (store information). These entities will contain all the critical information needed to make the restaurant's supply chain system information well-stored and easy to modify. Restaurant administrators can make this restaurant management system meet individual needs by updating key attributes.

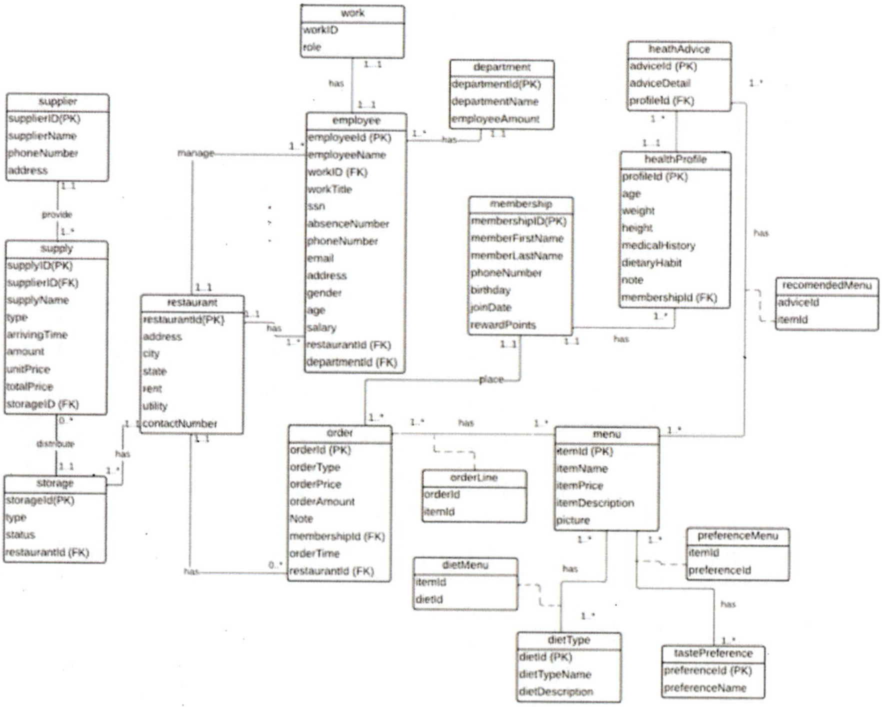

Fig. 2 ER diagram of database

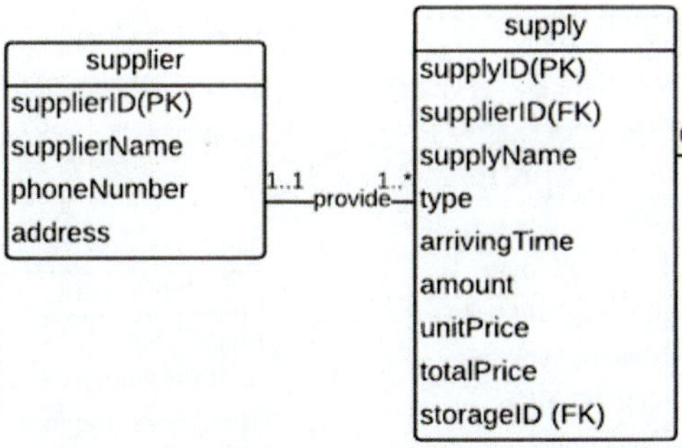

Fig. 3 ER diagram of supply chain

(2) *Employee Information Management*

In the restaurant employee information management part, this ERD further improves the labor management through two key entities: employee and departments (shown in Fig. 4). Under the employee entity, the basic information of restaurant employees will be stored, and the attribution of different employees will be distinguished by establishing relationships with the restaurant table and the department table. The restaurants table stores the information about certain restaurants using this restaurant management system. The department table further divides employee information, which is convenient for restaurant managers to retrieve employee information and adjust laboring.

(3) *Order and Menu Information System*

The menu table (shown in Fig. 5) is designed to store the food information of the restaurant, and the manager can easily modify the elements of the products. The details of the orders accepted by the restaurant will be stored in the order table, and a bridge table orderline will be used to establish a relationship with the menu table. The orderline table will store the information of the meals selected by the customer. This part is very critical because the function of providing dietary nutrition and health advice to customers will be derived through the two tables order and menu.

The menu information management part (shown in Fig. 5) includes 3 main entities. The first entity is the menu entity which could store the information of all the items the restaurant offered for customers to purchase. Businesses can add or modify information details such as item prices to items in the menu. Beside the menu entity, there is one entity called "dietType" which includes many kinds of diet like keto diet for instance. This entity can classify items in the Menu Entity. For example, some foods that fit the Keto Diet are grouped together. This makes it easier for people to order when they look at the menu and choose the kind of food that suits them. There is also an entity called diet preference, which lists some special dietary preferences, such as liking spicy food or not accepting Onions. This preference information will help Menu classify. In this way, when people choose food, they can also get recommended dishes according to their selected dietary preferences.

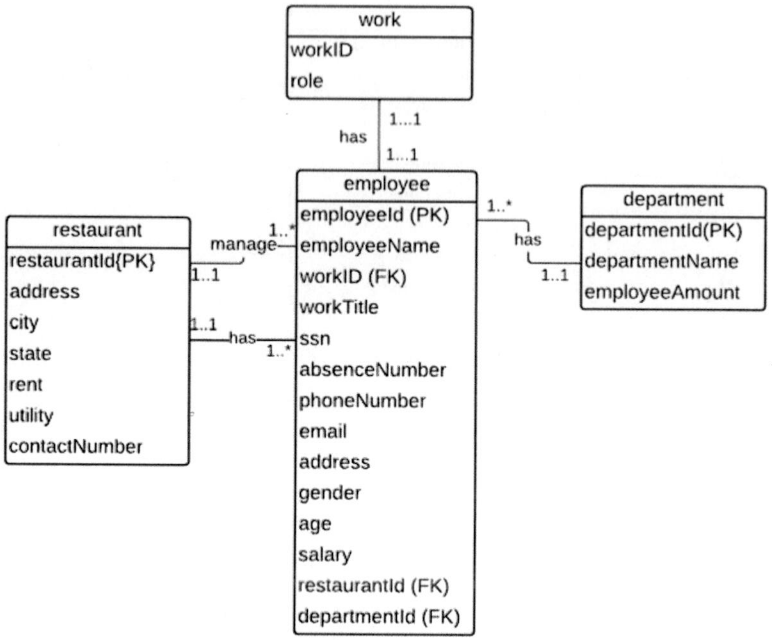

Fig. 4 ER diagram of employment

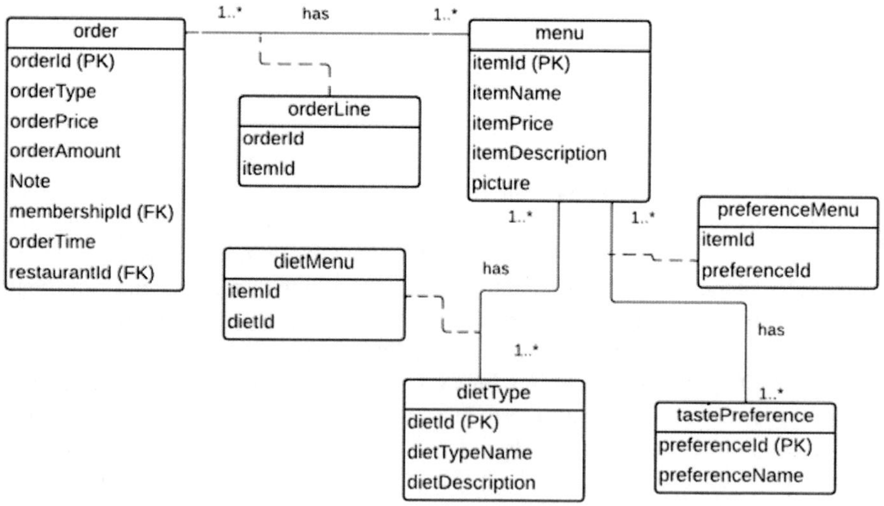

Fig. 5 ER diagram of order and menu

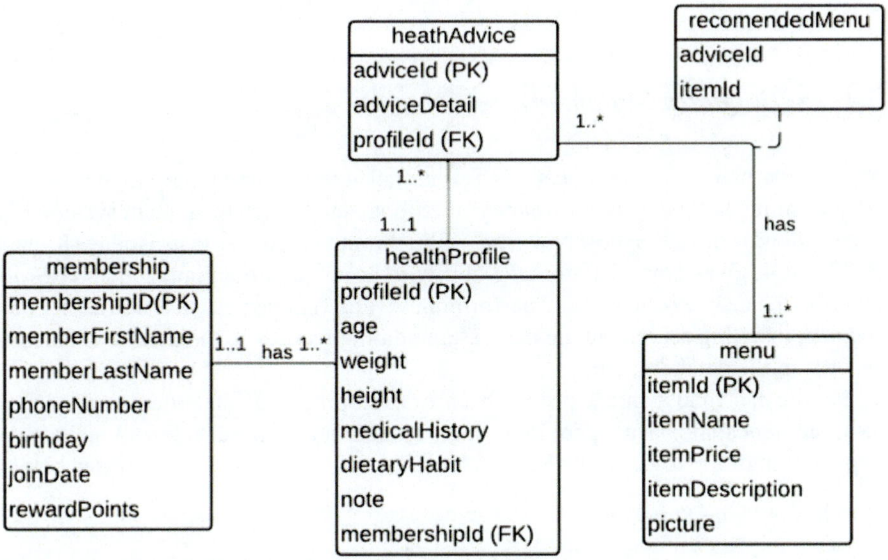

Fig. 6 ER diagram of Membership Information Management

(4) *Membership Information Management*

The membership information management part includes 2 main entities (shown in Fig. 6). The first entity is the membership entity. This includes the basic information of the registered members like membership id, member's name and the reward points. The other entity is the health profile entity. The relationship between this entity and the membership entity is the one-to-many relationship. The registered member could set up many profiles to get health advice for his or her family or friends. There is an attribute called "note" in this entity which could be used by the members to add notes and distinguish the profile they created.

The nutrition and health advice information management part (shown in Fig. 6) includes 1 entity which is the health advice entity. After members create the health profile and insert all the needed body information into the profile, the health advice will be given. Each health file corresponds to a health recommendation. Health recommendations may also change based on changes in data from health records. While providing the health advice, the customers will also be provided some recommended dishes from the menu in line with health advice. This makes it easier for customers to choose.

3 Implement

3.1 SQL Query Statement

SQL (Structured Query Language) is a relational algebraic and tuple-based relational programming language used to manage relational database management systems and is considered one of the most powerful programming languages in use today (Kanade 2022). SQL gives administrators the ability to access and manipulate databases for all critical data in a restaurant by performing several functions such as retrieving data records, inserting records in the database, updating records in the database, deleting records in the database, etc.

For the data management system that are developing, SQL statements will be used to implement some basic operations on the database, and this article will include the explanation of the logic of the SQL statements and some practical exercises below.

(1) *Insert Value Statement*

For the data management system to perform properly, the administrator needs to insert data into several key tables to make it work (shown in Fig. 7). The following SQL statements will insert values for restaurantId, location, city, state, zipCode, rent, utility, contactNumber into the restaurant table. In practical management scenarios, it will often encounter the need to import new data, add when the imported data does not exist in the database, and update when it already exists and needs to be modified (MySQL 2022). To make database management more efficient, it should follow the insert statement with the ON DUPLICATE KEY UPDATE clause. This is because SQL determines whether data is duplicated based on the primary key. If the inserted row has a duplicate value in the unique index or primary key of an existing record in the table, then the old row is updated; if the inserted row does not have a duplicate data in the unique index or primary key of an existing record in the table, then the new record is inserted.

Fig. 7 SQL insert query

```
select * from restaurant;
INSERT INTO `restaurant`
(`restaurantId`,`location`,`city`,`state`,`zipCode`,`rent`,`utility`
,`contactNumber`)
VALUES ('1', '1050 E Trinity Mills Rd', 'Carrollton', 'TX',
'75006', '11500 ', '1500 ', '2104776354')
on duplicate key update location=values(location),
city=values(city), state=values(state),
zipCode=values(zipCode),
rent=values(rent), utility=values(utility),
contactNumber=values(contactNumber);
select * from restaurant;
```

Fig. 8 SQL retrieve query

```
select healthProfile. *
from healthProfile
where membershipId = 10000;
```

Fig. 9 Retrieve data from two table

```
select menu. *
from menu
join `orderLine` on `orderLine`.menuId = menu.id
join restaruantHealth.order on `orderLine`.orderId =
restaruantHealth.order.id
where orderId = 1001;
```

(2) *Retrieve Data Statement*

In information systems, access to data is a very critical execution. It is important if managers want to retrieve a particular value from a large database. For example, if the user wanted to retrieve the health profile of a customer with a membership ID of 10,000, the following select statement could be applied (shown in Fig. 8).

If the user wants to get the orderline information and retrieve the items from the menu (shown in Fig. 9), the "join" statement should be used to get the information in the two tables at the same time.

(3) *Delete and Update Statement*

When the manager wants to update the data through this restaurant management system, he only needs to use the update statement to complete the operation (shown in Fig. 10). Also, some data needs to be arranged in a specific way for managers to refer to. For example, if the user wants to update the water and electricity situation of restaurant 1 and arrange the water and electricity of all restaurants in descending order to observe the water and electricity consumption of each restaurant, the following statement can be used with the desc command at the end to complete the descending order.

If a customer wants to cancel the membership, they need to use the delete statement to delete the membership-related information. Similarly, the delete statement can also be used in many other scenarios, such as deleting an entire row or an entire column (shown in Fig. 11).

Fig. 10 Update query

```
select * from restaurant;
update restaurant set `utility`=108 where restaurantId=1;
select * from restaurant order by `utility` desc;
```

Fig. 11 Delete query

```
deleteMembership
select * from membership;
delete from membership where membershipId = 10000;

delete a restaurant row
select * from restaurant;
delete from restaurant where restaurantId = 1;
select * from restaurant;
```

3.2 Front End Implement

In order to further implement the front-end of the database, this paper shows s Login page of the website and the health profile page of membership, and the menu page sorted by diet type.

(1) *Login Page for All Users*

First, the login page is a door that users must open in order to take full advantage of the site's user experience. This is the starting point for people to browse the site, so it must be there. This section is designed with pictures related to healthy eating and relatively simple text (Shown in Fig. 12).

(2) *Health Profile Page*

This part is also designed in a simple style. It is desired to keep the fonts on the pages consistent, and the pages brief. The webpage doesn't need to have too many elements, otherwise the user will get confused. But web pages need to have basic functions as well as a brief introduction (Liu and Ma 2011; Joo 2017). This page has all the data needed about the user's body and health. Every part except the "Note" part must be filled in, just as all information is required to be not null when designing the entity of this part. All the information should be prepared before providing users further health and diet advice. The "Note" part is designed by us to be "not null". For the convenience of users, the relationship between the profile Entity and membership Entity is designed as a 1-to-many relationship, which means that a member can create multiple profiles. For example, users can create profiles of their family or friends to get advice. This "note" section is what the user chooses, so they can put in their relationship with this person or whatever content they want (Shown in Fig. 13).

(3) *Menu Sorted by Diet Type Page*

In this part, first, the style and color of the subject are kept the same. In order to enable users to better find their own food, this page provides selection according to diet type and taste preference. And the system also provides recommended meals

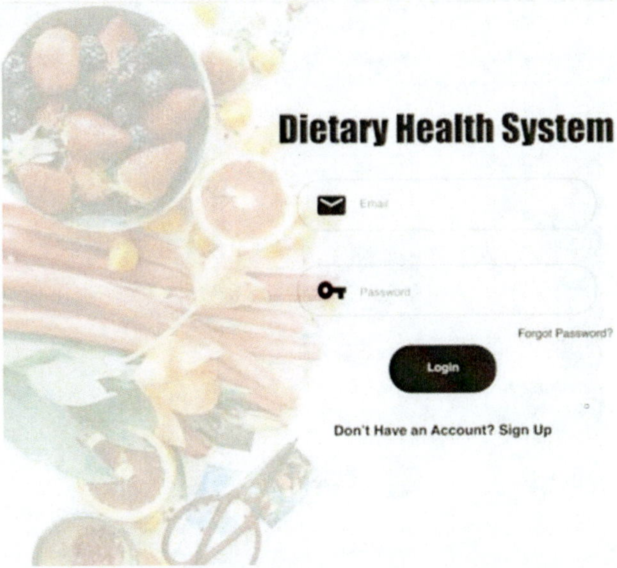

Fig. 12 Login page for all users

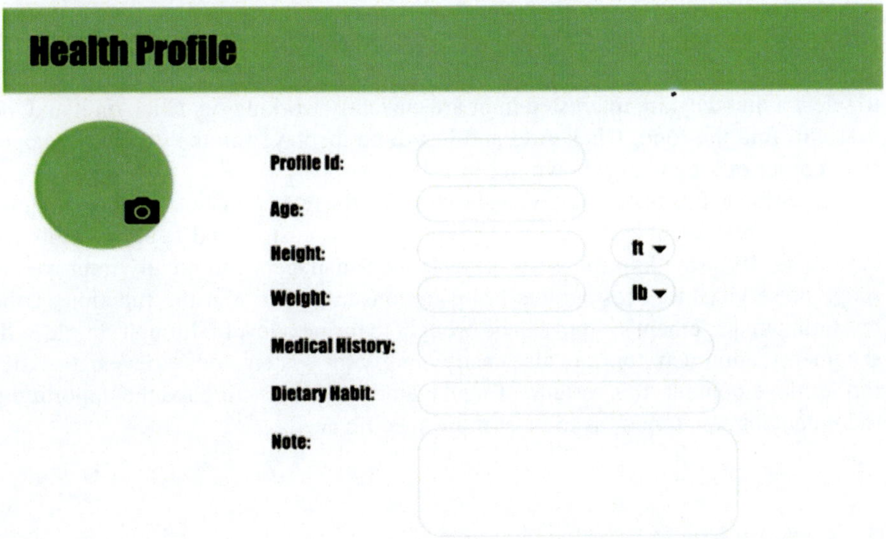

Fig. 13 Health profile page login page for users

Fig. 14 Menu sorted by diet type

after giving health advice. The example shown in the section of web design is the menu page classified by Diet Type. Users can swipe through the Diet Type module to select a diet they are interested in or are currently working on. Click on it and the user will find the food. Click Add and it will be displayed in the cart in the upper right corner of the menu (Shown in Fig. 14).

The eating habits of customers can be stored in the restaurant management system, so that the restaurant can launch personalized and customized healthy meals for customers. By providing this service, restaurant managers can greatly improve the scope of services that restaurants bring to consumers, so that the function of the restaurant management system can be brought to the next level. Through simple SQL statements, administrators can also easily modify the system and retrieve data. After further development, this restaurant management system will have the opportunity to implement new functions to meet more specific needs.

4 Conclusion

According to the research in this paper, the eating habits of customers can be stored in the restaurant management system, so that the restaurant can launch personalized and customized healthy meals for customers. By providing this service, restaurant managers can greatly improve the scope of services that restaurants bring to consumers, so that the function of the restaurant management system can be brought to the next level. This service can also improve customer expectations for the catering industry, as well as customer experience. Through simple SQL statements, administrators can also easily modify the system and retrieve data. For restaurant managers, this catering management system will consume less learning cost, make it easier for operators to get started, and at the same time incorporate new functions. After further development, this restaurant management system will have the opportunity to implement new functions to meet more specific needs. The author believes that the catering industry is an industry that directly serves people. In the future, catering services will be more closely integrated with people's living habits, and the management information system will help people's dietary health. In future research, the author will conduct research based on the research findings of this article and continue to explore more usable values of information systems to the catering industry, as well as more functions that have not yet been developed.

References

Aman F, Masood S (2020) How nutrition can help to fight against COVID-19 pandemic. Pak J Med Sci 36

Joo H (2017) A study on understanding of UI and UX, and understanding of design according to user interface change

Kanade (2022) What Is SQL? Definition, elements, examples, and uses in 2022

Kurniawan B, Zulfikar MF, Valentina T (2019) Developing restaurant information system to support decision making

Lifshitz F (2021) Nutrition and growth

Liu H, Ma F (2011) Research on visual elements of web UI design

MySQL (2022) MySQL: MySQL 5.7 reference manual: 13.2.5 INSERT statement

Revere D, Turner AM (2007) Understanding the information needs of public health practitioners: a literature review to inform design of an interactive digital knowledge management system

Wang J, Yeoh EK, Yung TK, WongMC, Dong D, Chen X, Chan MK, Wong EL, Wu Y, Guo Z, Wang Y, Zhao S, Chong KC (2021) Change in eating habits and physical activities before and during the COVID-19 pandemic in Hong Kong: a cross-sectional study via random telephone survey. J Int Soc Sports Nutr 18

World Health Organization (20210) Fact sheets: malnutrition

Research on Energy Saving Scene of 5G Base Stations Based on SOM + K-Means Two-Stage Clustering Algorithm

Jiahuan Zheng, Yong Xiang, and Siyao Li

Abstract This paper proposes a SOM + Kmeans two-stage clustering algorithm to adaptively cluster the daily load curve of 5G base stations and use silhouette coefficients to select the best clustering results, in order to maximize the energy-saving space of the base station and solve the problem of intelligent discovery of green base stations in the whole network. The daily tidal phenomenon and energy-saving periods are further analyzed to identify the energy-saving scene, and then a differentiated energy-saving strategy is periodically adopted for the base station to maximize network utilization and achieve the effect of smart energy-saving. Experimental analysis proves the efficiency and effectiveness of the algorithm. It is estimated that the energy- saving can reach about 15%, which can be applied to the intelligent management of 5G base station energy consumption to improve the energy efficiency of 5G networks.

1 Introduction

Single 5G base station has a lower coverage range, covers the same region, and requires more base stations than a 4G network, according to the spectrum resources used by domestic 5G networks. It will be extremely difficult for operators to keep their costs down on account of 5G base stations predicted to use 3–5 times as much power as 4G (Sun et al. 2022a). Therefore, it is crucial to initiate research on base station energy saving and consumption reduction (Erman et al. 2006).

J. Zheng (✉) · Y. Xiang · S. Li
Research Institute of China Telecom Co., Ltd., Guangzhou, China
e-mail: zhengjh@chinatelecom.cn

Y. Xiang
e-mail: xiangy11@chinatelecom.cn

S. Li
e-mail: lisy38@chinatelecom.cn

S. Yadav et al. (eds.), *Proceedings of 2nd International Conference on Artificial Intelligence, Robotics, and Communication*, Lecture Notes in Electrical Engineering 1063, https://doi.org/10.1007/978-981-99-4554-2_25

Scholars both domestically and internationally have conducted extensive research work in response to the issue of the base station's energy-saving scenario and have produced positive outcomes. The classification of BTS energy-saving scenarios can be separated into two groups based on the modeling approach: the first category is based on simple linear models or expert experience, relying on human experience to judge and operate; The second is based on machine learning algorithms, such as gradient lifting regression tree, neural network, etc. The experiment indicates that the machine learning algorithm shows better on energy-saving compared with traditional methods.

Time series clustering algorithm is an effective method for mining the similarity between time series, which can find some potential patterns from data. In recent years, many scholars have applied clustering algorithms to user behavior, business analysis and other aspects, and achieved superior results. Erman et al. (2006) and others used K-Means and DBSCAN algorithms to effectively identify traffic groups with similar communication characteristics; Zhu et al. (2014) and others found the emotional characteristics of different groups by clustering the time series data of microblog users. Xiaodi et al. (2019) proposed a temporal clustering method for user Internet behavior based on symmetric KL distance; Xiangjian (2019) proposed the application of k-shape algorithm based on morphological clustering of load curve in power load.

The aforementioned techniques are susceptible to beginning values and are prone to local optimization. The existing method on energy-saving scenarios is comparatively inflexible, and there are currently limited clustering analysis literatures including BTS side data. For this reason, this paper proposes a two-stage clustering algorithm based on SOM + Kmeans, combined with the iterative correction of silhouette coefficients, which can adaptively cluster the daily load curve of BTS services without preset parameters. In consideration of the waste of network resources caused by "tidal effect", the scene is personalized based on not only the weekly effect but also energy-saving period, which show greatly performance in the energy-saving.

In the 5G + AI era, the network needs to achieve intelligent dynamic online energy conservation management. Using machine learning algorithms to model and train historical data, find the rules, identify typical network energy conservation scenarios, and automatically turn on or off the carrier, increase or decrease the Mass MIMO power amplifier (Wu et al. 2022b), RF channel, and other energy conservation methods according to the load status of the base station, it is possible to automatically collect network traffic/configuration information. To ensure automation without impacting network performance, the network performance metrics are tracked and forecast in real time.

2 Related Work

2.1 SOM Clustering Neural Network Algorithm

SOM (Vesanto and Alhoniemi 2000) is an unsupervised algorithm with strong learning ability. It can automatically classify input patterns. It mainly applies dimension reduction to map high-dimensional input to lower space which is to make decision. SOM is composed of a single-layer neural network, which includes two layers, the input layer and the output layer. The output layer is also called the computing layer or the competition layer. SOM learning consists of the competitive process, cooperative process, and adaptive process. First, the network converts multi-dimensional input data into discrete low-dimensional data, and then it is represented as a local area or active point in the network. After initialization, learn to compete, and then cooperate and adapt. Compared with the weight distance between the input data and all neurons after competition, the neuron with the smallest distance will be chosen; The cooperation process shows that the superior neurons have an influence on the adjacent neurons in the dominant neighborhood, guiding the adjacent neurons to align with the superior neurons. The adaptive process makes the winner's neurons and their neighboring neurons in the winning region more sensitive to specific input values and improves the response and excitability to subsequent input modes. The adjacent neurons close to the superior neurons are more suitable for inputting samples with the same characteristics than the adjacent neurons far from the superior neurons. It can be seen that the SOM network is a very ideal clustering method with simple structure, self-organization, self-learning ability, strong pattern recognition ability and ourstanding classification effect. Figure 1 shows the SOM network structure.

2.2 SOM + K-Means Two-Stage Clustering

The k-means algorithm is fundamental, its principle is straightforward, and its calculation is convenient and fast. It has high accuracy when the number of clusters and the centre point is known, but its shortcomings are also very obvious. For example, the

Fig. 1 LSTM network structure

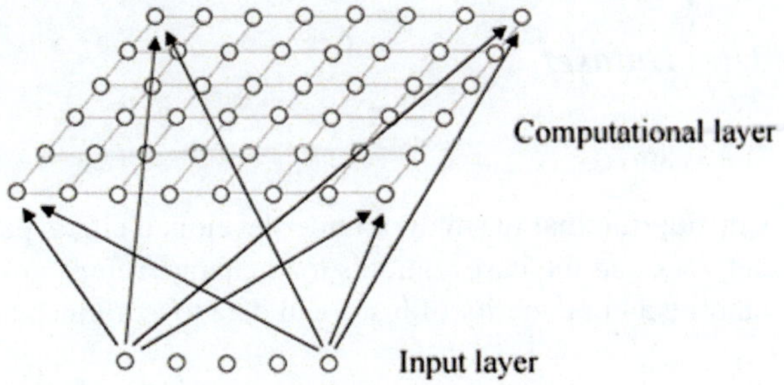

initial value is influenced greatly on the clustering results, falls into local optimization easily, and is sensitive to "noise" and isolated point data, which greatly limits its application scope and effect. SOM neural network has a strong anti-interference ability to noise data. It does not need to specify the cluster number and cluster centre during initial clustering. It can convert high-dimensional data, reduce dimensions and map to low-dimensional space. It can handle nonlinear data well, but it cannot give accurate and rich information on the cluster process.

The SOM + K-means two-stage clustering method is applied in this paper to combine the benefits of the aforementioned two methodologies. SOM performs initial clustering on enormous data samples in first. In the second stage, further clusters form the final clustering results by initializing the results of K-means as input. Similar feature vectors are considered to belong to the same category, and the number of type and the centre point of each type to be confirmed.

2.3 Silhouette Coefficients

Silhouette coefficients (Ogbuabor and Ugwoke 2018) is a method to evaluate the clustering result. It is used to compare the performance of different algorithms or algorithms with different parameters on the clustering results with the same data.

The silhouette coefficients of sample *i* is defined according to the degree of intra cluster and inter-cluster dissimilarity of sample *i*.

$$s(i) = \frac{b(i) - a(i)}{\max\{a(i), b(i)\}} \tag{1}$$

$$s(i) = \begin{cases} 1 - \frac{a(i)}{b(i)}, & a(i) < b(i) \\ 0, & a(i) = b(i) \\ \frac{a(i)}{b(i)} - 1, & a(i) > b(i) \end{cases} \tag{2}$$

3 Experiment and Result Analysis

3.1 Dataset

Data Source

Considering that in small or underdeveloped cities, people have low demand for the network and the base station's load change rule is not substantial, this article has to employ a big quantity of historical data to examine and analyse the scenarios where the base station might save energy. In order to confirm the viability and generalizability of the model, a large-scale, operational network service base station of an operator in a city is chosen as the research object.

Variable Description

This paper selects 10 consecutive weeks of time series data from an operator in a city in 2022, covering 24-h hourly service load data records of more than 60,000 base stations. The data types are as follows:

(1) Equipment data. Including the manufacturer, room division, frequency point, direction angle, etc.
(2) Business data. This includes the active user, the average upstream and downstream traffic of the PDCP (Packet Data Convergence Protocol), and so on.
(3) Network data. It includes the utilization ratio of uplink and downlink PRB (Physical Resource Block) and the occupancy ratio of various channels.

3.2 Result Analysis

Periodic Effect Identification of Base Station Traffic Load

In this paper, SOM + K-means two-stage clustering is used to identify the day-of-the-week effect types of base stations combined with silhouette coefficients. The three types are defined as follows:

- Obvious day-of-the-week effect: If the silhouette coefficients of the clustering result is greater than 0.65, it means that the clustering result on the daily load is better, evidently it has the obvious weekly effect.
- Consistent weekly trends: If the silhouette coefficients of the clustering results is less than 0.65 and the average similarity is greater than 0.9, it means that the base station has extremely similar load daily, that is, the base station has the same trend throughout the week.
- No obvious effect: Except for the above two categories, it is defined without obvious change rule.

Daily Tidal Phenomenon and Energy-Saving Period Identification

(1) In order to verify daily tidal phenomenon, the first step is to calculate the tidal effect coefficient of the base station.
Assuming that it is the daily traffic load curve function of one base station, the tidal coefficient is defined as:

$$\Delta = \max|f(x)| - \min|f(x)|, T \in [0, 1] \tag{3}$$

When the tidal coefficient is larger than 0.5, it is concluded that the base station has an obvious daily tidal phenomenon; otherwise, the daily tidal phenomenon is not obvious.

(2) The second step is to inference the heavy traffic period and energy-saving period for the base station with the daily tidal phenomenon, and further define the energy-saving scene of the base station.
According to the tidal coefficient, define the continuous time point interval that corresponds to $f(x) < \frac{1}{4}\Delta$ as the energy-saving period, so that the daily load curve of the base station is divided, and multiple energy saving periods with energy-saving value are selected. Further targeted energy-saving strategies are adopted, such as carrier wave shutdown by time period, etc.

Intelligent Recognition of Personalized Energy-Saving Scenarios

Combined with the weekly effect type and daily tidal phenomenon of the base station, multiple regular energy-saving periods in a day can be accurately identified as the personalized energy-saving scenes:

For base stations with obvious weekly effect, they can be grouped under two or three-modes according to the number of categories of daily classification, (others are not considered). Additionally, based on the number of days, the two-mode base stations are divided into effects for 1 + 6, 2 + 5, and 3 + 4 days of the week. The 1 + 6 day-of-the-week effect represents that the traffic load curve of the base station can be divided into two types in seven days a week: the first category is that of a single day, while the second category is that of a combination of six days (such as the first category is Sunday, and the second category is Monday to Saturday). For the three-mode base stations, it can be divided into 1 + 1 + 5, 1 + 2 + 4 and other day-of-the-week effects.

Under the day-of-the-week effect, the tidal phenomena of different types are further subdivided. The two-mode base stations correspond to the energy-saving periods of Class 1 and Class 2, respectively, and the three-mode base stations correspond to the energy-saving periods of Class 1, Class 2 and Class 3, respectively, so as to describe the personalized energy-saving scenes of the base station more accurately and in detail. Individual energy-saving scenes for base stations with the consistent weekly trend, that is, with similar business trends seven days a week, can only be defined in accordance with the first type of energy-saving period.

Three scenes are shrewdly chosen as examples to implement various energy-saving tactics based on the description of customized energy-saving scenes provided above.

Base Station with Obvious Day-of-the-Week Effect

Figure 2 manifests the traffic load changes of a base station in a commercial building in Guangzhou (ID: 856338) for ten weeks from 2022-06-03 to 2022-08-12. It can be seen from Fig. 2a that this type of base station repeats a similar fluctuation law every week, that is, the load in the early stage of the week is relatively high, and the load is at a trough at the weekend.

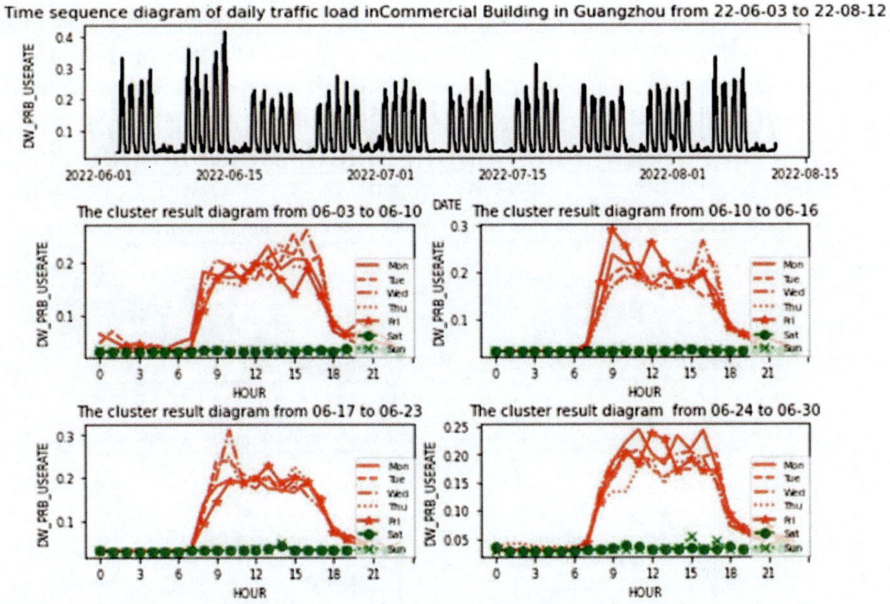

Fig. 2 SOM-KEAMS clustering result on BST-ID (ID: 856338)

The business mode of the base station can be divided into two categories, that is, the load curve is roughly similar from Monday to Friday, and the load trend at the end of the week is roughly the same. It belongs to the base station with obvious day-of-the-week effect. Considering that the business load of commercial office base stations on weekdays is significantly higher than that on weekends, the energy-saving period is from 0:00 to 7:00 in the morning; while employees rest at weekends and all time energy can be saved. This result conforms to the actual situation and therefore is of positive significance to the enterprise energy-saving. Therefore, the personalized energy-saving scene of the base station is "5 + 2 day-of-the-week effect (working day effect): 0–7 o'clock on weekdays, all day on weekends".

Base Stations with Consistent Weekly Trends

Figure 3 demonstrates the business load changes of the base station named Xintangying Industrial Park (ID: 485435) in ten weeks from 2022-06-03 to 2022-08-12. The daily business load curve of this type of base station is roughly the similar, and it is at a low business peak in the early morning. Through analysis on energy-saving period, the personalized energy-saving scene of the base station is identified as "weekly trend consistent: 0–7 o'clock every day".

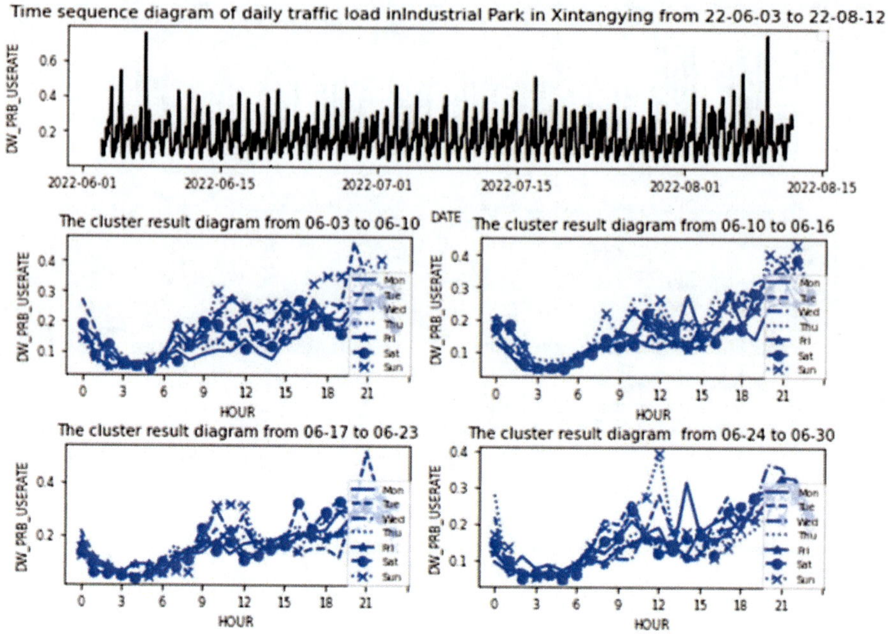

Fig. 3 SOM-KEAMS clustering result on BST-ID (ID: 485435)

4 Irregular Base Station

Figure 4 displays the traffic load changes of the base station named Huashan Sub-branch (ID: 485710) for ten weeks from 2022-06-03 to 202-08-12. It can be seen that the daily traffic load of this type of base station fluctuates randomly. It is not suitable to implement the energy-saving strategy, so as not to impact the user experience.

Energy-Saving Benefit Assessment and Analysis

Based on the SOM + K-means clustering and daily tidal analysis, the personalized energy-saving scenes of the base station can be quickly and effectively identified, so that the energy-saving strategy can be selected periodically and intelligently as shown in Table 1.

From the results, the personalized energy-saving scenes of a commercial building base station in Guangzhou (ID: 856338) is a 5 + 2 working-day effect. Class 1 (Monday to Friday) energy-saving period is 0:00–8:00 and 19:00–24:00. The energy-saving period of Class 2 (weekend) is 0:00–24:00, and the energy-saving time can last up to 108 h a week. Two strategies a week can be adopted and implement the appropriate energy-efficiency measures respectively in the two modes. Up to 64% of the energy used by base stations can be saved.

The personalized energy-saving scene of Base Station named Xintangying Industrial Park (ID: 485435) belongs to the consistent weekly trend. The energy-saving

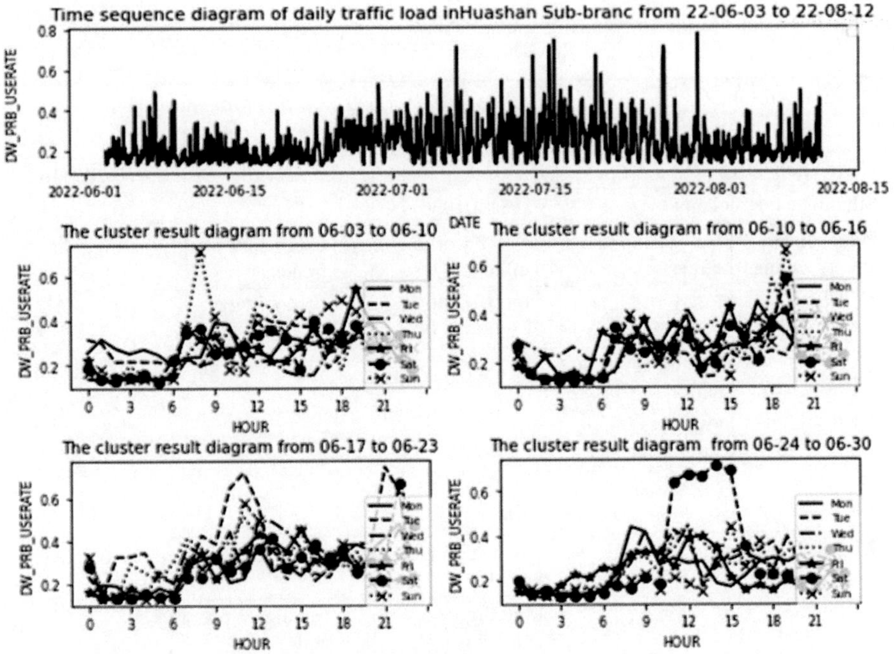

Fig. 4 SOM-KEAMS clustering result on BST-ID (ID: 485710)

period of a single day is from 1 to 7 o'clock. The strategy can be implemented once a week, and the energy-saving time can be 42 h per week. It is estimated that base station energy consumption can be economized by up to 25%. As for the base station of Huashan Branch Office (ID: 485710), it is not appropriate to conduct energy-saving methods to assure user experience since this type of base station is susceptible to emergencies and random fluctuations.

The results illustrate that the intelligent and efficient identification of personalized energy-saving scenarios has significant guiding significance for energy consumption management. It can be monitored and estimated based on the data from all of the base stations in a city. The average power usage can be reduced by more than 20% when network performance quality is guaranteed, which has a substantial impact on energy conservation.

5 Conclusions

This project was undertaken to propose a SOM + K-means clustering algorithm, combined with the daily tidal effect and energy-saving period based on the daily traffic load curve.

Table 1 Personalized energy-saving scenarios for base stations

Base_station ID		856,338	485,435	485,710
Clustering results		Class 1: Mon–Fri	Class1: Mon–Sun	–
		Class 2: Sat–Sun		
Correlation coefficient		–	0.932719	0.289416
Silhouette coefficients		0.81401936	–	0.236171
Personalized energy saving scenarios	Day-of-the-week effect	5 + 2 working day effect	Consistent weekly trends	Irregular
	Energy-saving period in Class 1	0:00–08:00 19:00–24:00	01:00–07:00	–
	Energy-saving period in Class 2	0–24:00	–	–
	Energy-saving period in Class 3	–	–	–
	Energy saving duration per week	$12 \times 5 + 24 \times 2 =$ 108 h	$6 \times 7 = 42$ h	–
Weekly energy saving strategy		Two strategies a week	One strategy a week	Not suitable
		Intelligent opening and closing 0:00–07:00 and 19:00–24:00 from Mon to Fri 0:00–23:00 from Sat to Sun	Intelligent opening and closing 01:00–07:00 from Mon to Fri	
Energy saving mode		1. Symbol shutdown 2. Channel off 3. Deep sleep, etc.	1. Symbol shutdown 2. Channel off 3. Deep sleep, etc.	–
Maximum energy-saving space		64%	25%	–

An extensive empirical study manifests the reliability of the clustering effect and the validity of the algorithm for identifying energy-saving fields which helps operators improve the energy efficiency of base stations on the basis of meeting the requirements of service QoS (Qiang et al. 2019) and meet expectations of energy saving of 5G base stations. This insight gained from this study may be of assistance to realize intelligent management of 5G base station energy, use network resources effectively, and enhance operators' market sustainable development capabilities.

References

Erman J, Arlitt M, Mahanti A (2006) Traffic classification using clustering algorithms. In: Proceedings of the 2006 SIGCOMM workshop on mining network data, pp 281–286

Ogbuabor G, Ugwoke FN (2018) Clustering algorithm for a healthcare dataset using silhouette score value. Int J Comput Sci Inf Technol 10(2):27–37

Qiang X, Wei G, Yansheng C (2019) Research on the implementation of differentiated QoS policy in LTE network based on user value. Post Telecommun Des Technol 6:88–92

Sun Y, Fu C, Cui C, Wen Q (2022a) Research on operation control strategy of energy-saving power supply system of 5G communication base station. In: 2022a 4th international conference on communications, information system and computer engineering (CISCE), pp 130–133. https://doi.org/10.1109/CISCE55963.2022.9851047

Vesanto J, Alhoniemi E (2000) Clustering of the self-organizing map. IEEE Trans Neural Netw 11(3):586–600

Xiangjian Z (2019) Research and implementation of clustering based on user behavior time series. Beijing University of Posts and Telecommunications, Beijing

Xiaodi W, Junyong L, Youbo L et al (2019) Morphological clustering algorithm of typical load curve using adaptive segmentation aggregation approximation. Autom Electric Power Syst 43(1):110–121

Wu T, Yuan L, Zhou A (2022b) Antenna selection technology research in massive MIMO system. In: 2022b IEEE Asia-pacific conference on image processing, electronics and computers (IPEC), 2022b, pp 967–970. https://doi.org/10.1109/IPEC54454.2022.9777612

Zhu J, Bai W, Wu B (2014) Social network users clustering based on multivariate time series of emotional behavior. J China Univ Posts Telecommun 21(2):21–31

Smart Substation Synthetical Smart Prevent Mishandling System Based on Topology Model and Intelligent IOT

Fan Zhang, Wenping Li, Xiaolong Tang, Juan Yu, Jiangshen Long, and He Xu

Abstract Aiming at the prominent problems of the current variable prevent mishandling system, such as the hidden problem of universal key, offline prevent mishandling mode and logic defects, this paper researches the smart substation synthetical smart prevent mishandling system, by firstly constructing a new prevent mishandling judgment model based on the topology analysis algorithm based on the knowledge representation method of prevent mishandling logic rules and logic judgment method. Secondly, relying on the intelligent IOT technology and wireless digital energy homologation technology, the synthetical smart prevent mishandling system composed of a smart prevent mishandling host, prevents mishandling the edge agent device, intelligent smart key, preventing the mishandling of the locking device, intelligent grounding wire module and other devices constructed to realize the integration and cooperation of multiple systems such as five preventions, lock control and access control of substations and the integrated error prevention function under multiple business scenarios. Finally, through the pilot application project, we will verify the feasibility of the architecture, process and function, and build a station-wide and process-wide synthetical smart prevent mishandling system for substations.

1 Introduction

1.1 A Subsection Sample

For special complex operations, emergency accident handling and other situations, it is necessary to configure and use the universal key to unlock the substation in order to carry on the operation. Once the universal unlocking key is used, the safety gatekeeper of the prevent mishandling rules is lost, and the "technical prevention" measures that should function are returned to "human prevention" (Zhao et al. 2022); The logic

F. Zhang · W. Li (✉) · X. Tang · J. Yu · J. Long · H. Xu
Anhui Nanrui Jiyuan Electric Power System Tech Co., Ltd., Hefei, China
e-mail: 1115379648@qq.com

S. Yadav et al. (eds.), *Proceedings of 2nd International Conference on Artificial Intelligence, Robotics, and Communication*, Lecture Notes in Electrical Engineering 1063, https://doi.org/10.1007/978-981-99-4554-2_26

rules of the current prevent mishandling system mainly use the logical operation of “with, or, not” of the telecommunication of the equipment’s “split-close” status to make judgments, in fact, in some cases the equipment in the split-close state may also be charged and cause danger, and the prevent mishandling logic method has inherent defects (Chao and Cheng 2021; Dang 2019). The problem of defective mechanical locks (Naibo and Qisheng 2019; Zhang et al. 2021), the problem of temporary grounding wire hookup management (Jinkui 2018).

Based on the knowledge representation method of prevent mishandling logic rules and logic judgment method, a new prevent mishandling judgment model based on topology analysis algorithm is constructed. Relying on IOT technology and wireless digital technology , solving the wireless communication and power supply problems of prevent mishandling devices and realize real-time online and status feedback of error prevention devices. Based on the above research, the research and development of synthetical smart prevent mishandling system of the substation will be carried out to realize the integration and cooperation of multiple systems such as five preventions, lock control and access control of the substation and the comprehensive error prevention function under multiple business scenarios.

2 Synthetical Smart Prevent Mishandling Device Based on IOT Communication Technology

Conduct research on the application of LoRa technology in the synthetical smart prevent mishandling system of smart substation, combines the access requirements of various scenarios of prevent mishandling logic to real-time online information, explores the wireless networking mode applicable to substation scenarios, and establishes a sensing network covering the whole substation; adopts LoRa communication for real-time online lock open/close state recognition and the single-step prevent mishandling verification interaction method to realize reliable data access of new smart prevent mishandling devices. The application technology of wireless SWIPT communication technology in smart keys and error-proof locks is used to realize the real-time identification and timely transmission of the opening; the IOT communication module is used to realize the sensing of the operating status of new smart prevent mishandling devices such as intelligent locks, smart keys, intelligent grounding wire modules and switchgear intelligent bolts, flow chart of topology analysis as shown in Fig. 1.

The overall architecture of the synthetical smart prevent mishandling system contains host, prevents mishandling edge agent device, smart key, prevents mishandling locking device, intelligent grounding wire module and other equipment, and the system architecture is shown in Fig. 2.

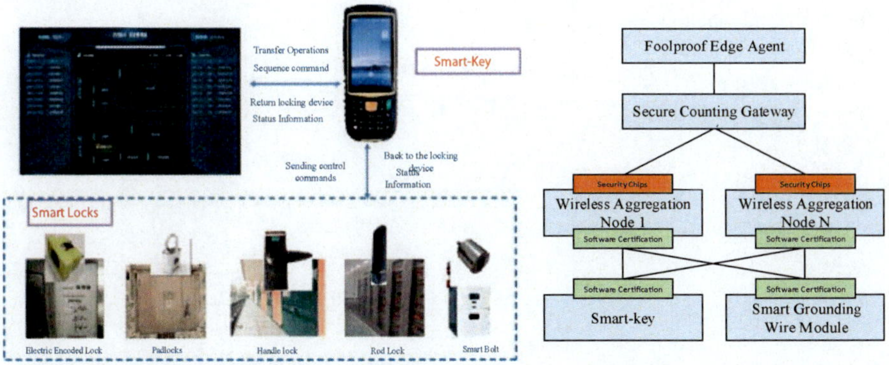

Fig. 1 Flow chart of topology analysis

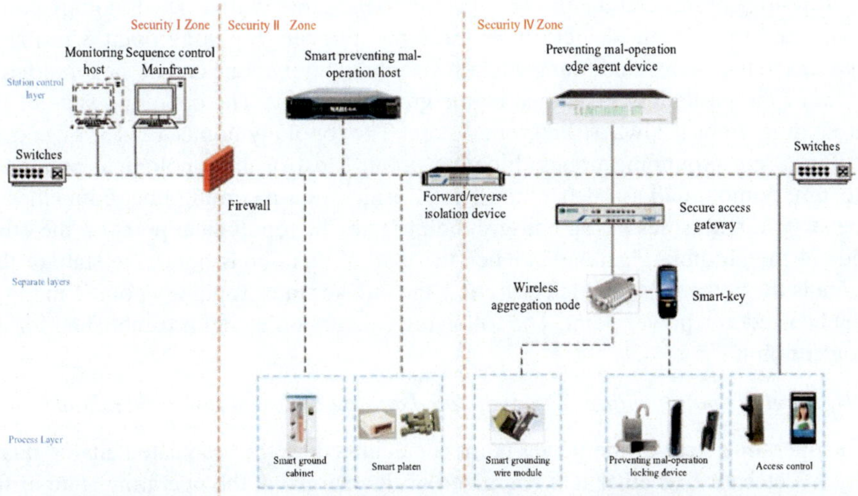

Fig. 2 Synthetical smart prevent mishandling model system architecture diagram

3 Synthetical Smart Prevent Mishandling Model Based on Topological Analysis Algorithm

In order to model the topology calculation, the model definition of IEC61850 needs to be extended by adding a definition of the type of topology points. The following model definitions for topology points are available.

nodal{topology point code. Topology point type. Operational state;}

There are three types of topological point definitions: power point, grounding point, and floating point. The topological point type owned by the grounding stake used to hang the grounding wire is the grounding point.

Fig. 3 Typical topological diagram

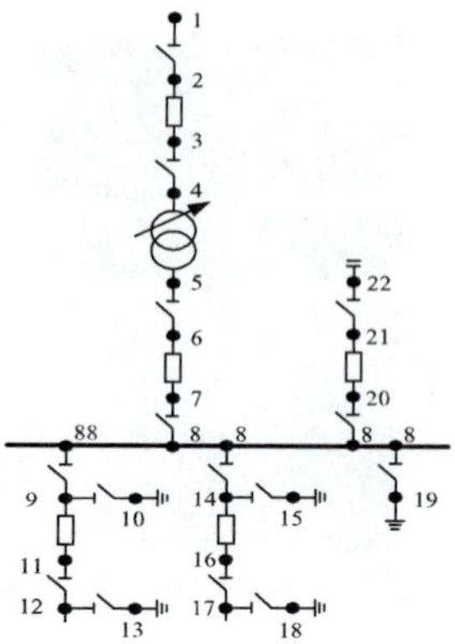

Topology points 2 and 3 in Fig. 1 belong to the same switch. The topology points connected to different devices have the same number. Topology point 8 in Fig. 1 belongs to the busbar and four switches. Grounding point: one end of the topological point of the grounding gate must be the grounded point. The operating state of the grounding point is always the ground state. The topology points 12, 13, 17 and 18 in Fig. 3 are grounding points. Floating point: Most of the topological points are floating points, such as busbar, main transformer, non-incoming line, both sides of the switch, both sides of the non-grounding gate, the topological point on the other side of the grounding gate, etc. When the load of this line is not 0, the state of this point is the powered state, otherwise it is the outage state. Topology point 1 in Fig. 1 can be used as a power point. The following calculation model is established for the power point.

PN{Topology point coding. The associated analog coding. Analog threshold}

The operating state of the power point is calculated by the associated analog quantity to determine whether it is energized or de-energized, the operating state of the grounding point is always grounded, and the operating state of the floating point is determined by the operating state of the topology island where it is located. The topology island generation algorithm is shown in Fig. 4.

4 Experimental Analysis

In this paper, the electrical main wiring of the substation shown in Fig. 3 is used as an example, and the system start state information is shown in Table 1. After the ground switch is opened, the system is initialized and analyzed using the topology analysis method, and the results are shown in Table 2.

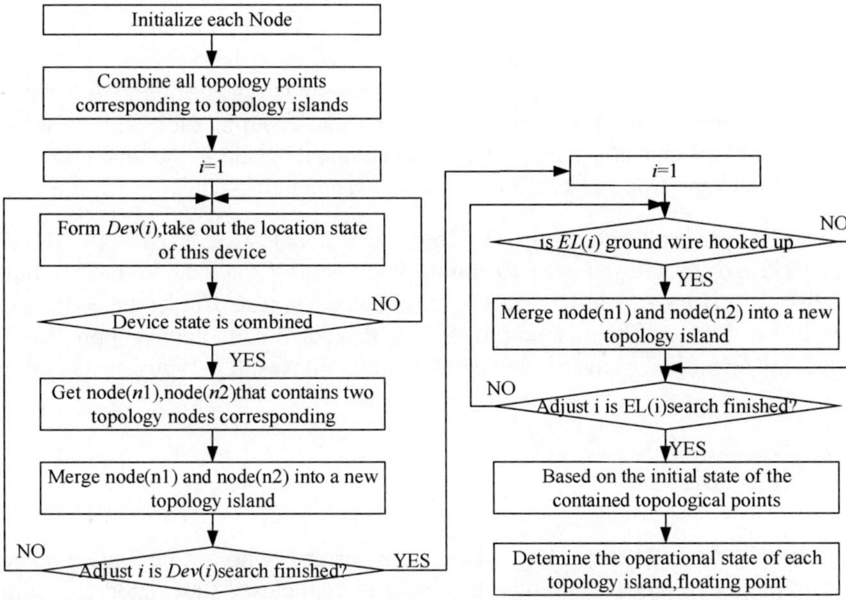

Fig. 4 Topology island generation algorithm calculation steps diagram

Table 1 Topological diagram initial information statistics table

Nodal type	Nodal number	Initial state
Power nodal	1	Charged state
Load nodal	12, 17	Suspension potential
Grounding nodal	10, 13, 15, 18, 19	Grounding state
Other nodal	Other nodals	Suspension state

Table 2 Division of electrical traffic circle assuming disconnect switch disconnection

Electrical traffic circle	Nodal number	Loop status
1	1, 2, 3, 4, 5, 6, 7	Charged state
2	10	Grounded
3	13	Grounded
4	15	Grounded
5	18	Grounded
6	29	Grounded
7	Other nodals	Power failure

(1) If the ground switch between nodals 14 and 15 is closed, the charged state of the electrical branch in which nodals 14 and 17 are located must be obtained separately, where nodal 14 belongs to electrical branch 1 and is in the charged

state; nodal 15, which belongs to electrical branch 4, is grounded and therefore cannot be operated.

(2) If you now want to disconnect the disconnect switch between nodals 7 and 8. The updated electrical circuit is shown in Table 2. From Table 2, we can see that after disconnecting the switch between nodals 7 and 8, the load nodals 12 and 17 change from the charging state to disconnecting state.

It can be seen from the results in Table 2 that nodals 8, 9, 11, 12, 14, 16, 17, 19, 20, 21, and 22 will fail after assuming that nodals 12 and 17 in Table 1 have been disconnected, so nodals 12 and 17 cannot be disconnected under the nodal state conditions in Table 1. The method proposed in this paper can effectively predict the possible misoperation in advance and give the prediction result to avoid misoperation.

5 Conclusion

(1) Topology analysis technology, change the current "with, or, not" logic operation judgment mode that only relies on "split and close" state, integrate multi-dimensional real-time data as a criterion, using topology to prevent mishandling technology to solve the traditional prevent mishandling cannot determine "whether the power is charged", "whether there is an obvious disconnection point", "whether there is grounding", "whether it will cause a large area blackout" and other logic defects.

(2) Introduce intelligent IOT technology to upgrade the IOT and intelligence of prevent mishandling devices such as locks, keys and ground wires in substations, realize the identification of the opening and closing status of prevent mishandling locks, online single-step prevent mishandling verification, real-time interaction of ground wire hook-up status and location information, and support the real-time online prevent mishandling function of the synthetical smart prevent mishandling system.

References

Chao Y, Cheng R (2021) Design of intelligent control system for networked dispatching of distribution networks based on topology error prevention. Electron Des Eng 29(19):146–149+154

Dang S (2019) Remote operation safety and error prevention technology for smart grid dispatch control system. Digit Technol Appl 37(6):119–121

Jinkui H (2018) R & D and application of anti-misoperation lockout logic rule generation and checking system based on typical connection. Electron Compon Inf Technol 16:5–9

Naibo S, Qisheng Y (2019) On the application of portable dive-proof system. Mech Electr Inf 6:21–22

Zhang H, Zhang S, Liu D et al (2021) Design and application of a visualization check system for anti-misoperation blocking logic in a substation. Power Syst Prot Control 49(12):181–187

Zhao J, Zhu H, Chen N et al (2022) Research and application of a new fault prevention system for inverted gate operation based on topological fault prevention model and IOT technology. China Plant Eng 8:69–71

Fresh Products Application Information Management System

Xueyan Yang

Abstract There has been a pronounced increase in online shopping since the start of the COVID-19 pandemic. Online supermarkets are one of the critical topics of research today. Today there is a wide range of online supermarket apps available nationally and internationally, and good progress has been made in transporting goods. However, there are still research gaps in managing food waste and fresh produce. Therefore, the research theme of this paper is to establish an information management system for a new production software. The methodology of this paper is as follows: firstly, the author analyses the functional requirements of the system, secondly the structural framework of the system is designed based on the requirements analysis, then the ER diagram is completed based on the understood data requirements, processing requirements, security and integrity requirements of the users and administrators, the database is created, tables are created, data is inserted, and finally the desired query functions are achieved through MySQL. Finally, the web design for the background management of product additions, user registration accounts, and login accounts was conducted by compiling HTML and CSS in VScode software. This paper has achieved the target query function. However, the query database steps are still slightly complicated, not convenient and fast enough, and the operational requirements of the administrator are also relatively high.

1 Introduction

With the continuous development and improvement of computer, communication, network, and large-scale database technology, many chain supermarkets choose to sell their goods on the Internet by using public or self-created shopping apps. In comparison, fresh products are high-frequency consumer goods in supermarkets. During the epidemic, consumers' online new food consumption demand has expanded, which has promoted the increase in the scale and proportion of online

X. Yang (✉)
College of Computer Science and Technology, Huai Bei Normal University, Huai Bei, China
e-mail: bbyxy@poers.edu.pl

S. Yadav et al. (eds.), *Proceedings of 2nd International Conference on Artificial Intelligence, Robotics, and Communication*, Lecture Notes in Electrical Engineering 1063, https://doi.org/10.1007/978-981-99-4554-2_27

fresh food retail, and at the same time, helped the steady development of the new food e-commerce industry. Affected by the epidemic, traditional grocery shopping consumption scenarios such as offline supermarkets and vegetable markets have been restricted, and consumers' online grocery shopping frequency has increased because this non-contact shopping method dramatically reduces the probability of catching the virus. Meanwhile, supermarkets, as a low-margin industry, can reduce production costs in the best way, and cut down the cost of procurement, the cost of distribution, and the cost of inventory management, which can help them to attract more customers at lower prices (Failed 2009).

There has been a pronounced increase in online shopping since the start of the COVID-19 pandemic. Barbara Baarsma and Jesse Groenewegen analyze the link between the pandemic and online grocery shopping in more detail. They distinguish between the impact of the local virus situation and the impact and public perception of the national virus situation. After that, they conclude that the pandemic has indeed played a role in promoting the development of online shopping (Baarsma and Groenewegen 2021). Online retail formats like fresh food e-commerce and store-to-home have proliferated since this year, according to data from the National Bureau of Statistics of China. Online shopping has continued to proliferate. The national online retail sales of physical goods increased from January to July by 15.7% yearly, 1.4% points more than they did from January to June. These sales made up 25.0% of all retail sales of consumer goods, an increase of 5.6% points over the same time last year.

The business model of Jingdong Home is a fresh food supermarket O2O platform model, which mainly provides offline users with online shopping services from offline merchants (Chao 2020). Consumers can purchase products from offline merchants based on a specific range of user locations through the Jingdong Home platform. The merchants can choose to provide home delivery services through their delivery or crowd-sourced delivery on the forum. Consumers can view, place orders, make payments and carry out after-sales services on the online platform. At the same time, the Jingdong Home platform can effectively connect traditional brand owners, retailers, logistics providers, and consumer resources, making online retailing more efficient and convenient. However, some scholars have used the questionnaire method to investigate the three main aspects of JD.com's users on platform service, merchant service, and logistics service, collecting data in the form of a five-star Likert scale and conducting factor analysis on the questionnaire results (Chao 2020). It was concluded that JD.com's current development problems are mainly reflected in product quality and delivery service, after-sales service, and shortage of goods (Chao 2020).

There are also many similar applications abroad, such as the mobile online shopping system for the uchumi supermarket. Utilizing eclipse, PHP, and documentation created in Microsoft Word 2007, the uchumi supermarket's mobile internet shopping system was implemented (Taban and Bakulembeze 2013). The shopping system already has the basics in place, and the user can use the mobile online shopping application, which displays information about the items that the shopping system has. The user can then add the selection of what they want to shop for to their shopping cart. After that, the user can always check the shopping list to see how much they need to pay. Finally, when the user chooses to order, they need to fill in their personal information in the order form. The management will receive this information to contact the user to inform them when the goods will be delivered.

Today, every business is faced with two challenges. It is improving efficiency and reducing costs. As industries continue to face increased global competition, the need for top-notch products, quality customer service, and guaranteed low prices is growing stronger. To compete successfully in today's marketplace, obtaining a fully functional data management system to support the business is critical to successfully managing these challenges and maintaining a competitive edge (Mkoba 2023). The two largest physical supermarket companies in Australia, Woolworths, and Coles, compete fiercely for customers' attention both in brick-and-mortar locations and online. ShopFast is the primary online rival and lacks a physical supermarket location to continue its online business (Freeman et al. 2003).

Nevertheless, some problems have not been dealt with, so many kinds of online supermarket platforms or software exist. A fundamental problem with online supermarkets is poor food management. Usually, this causes food to go wrong and end up in the garbage, costing much money and the environment, or it causes people to eat it and get sick (Aydin et al. 2017), especially for preserving fresh food, which is problematic. Hence, the problems of food safety and its waste are severe. The development of Fresh food online supermarkets will solve the problem (Xiang et al. 2014).

Moreover, it was found that online supermarkets must establish confidence with repeat online consumers if they desire to increase their online supermarket's electronic word of mouth (e-WOM) (Eliaz et al. 2021). The management of shelf life is therefore critical, not only in terms of the disposal of expired products but also in terms of whether customers will buy expired products or whether they will be dissatisfied with the supermarket because they are purchasing expired products at the same price as fresh products. Firstly, supermarkets can discount sales by stating in the product description that the product is about to expire. Secondly, product reviews could be opened to be visible to each customer when purchasing a product.

Convenience and time savings are the two most significant benefits of online shopping. At the same time, the danger of improperly valuing some products and concern about the choice and handling of perishables like meat, eggs, and vegetables are the most significant concerns for customers (Hanus 2016). Based on these, the study aims to design a new product application information management system with improved product information management functions and a more rational approach to managing fresh produce, which has a short shelf life. The customers can add any products they like into their online shopping carts from the goods display page and remove any unwanted items from shopping carts. They can browse the content in their shopping cart at any time, and the content will not be modified by anyone other than themselves. Until they submit the current order, the content in the shopping cart will be presented as the order content, and finally, the current cart will be emptied. They can then use their credit card to pay for the order and wait for the goods to be delivered. At last, they can evaluate products on the app.

While application administrators can update this database by adding new species of goods into it, after they analyze their sales volume and compare the purchase volume of customers of different commodities, application administrators can know which goods sell best through the query statement and then increase the purchased quantity of this kind of goods at the next purchase. Accordingly, they can also reduce purchasing goods that are not selling well. The important thing is that they can use this database to maintain goods better. They can classify different foods according to the temperature suitable for their freshness and keep them fresh. For the products about to expire, they can put their status pictures on the home page for scrolling at a low price to attract people to buy them, which can also avoid unnecessary waste.

The rest of this paper is organized as follows: Sect. 2 provides an overview of the structural design of new products application information management system. In Sect. 3, the author presents the database design of new products application information management system. Section 4 uses specific examples to demonstrate the main query functions. This section introduces the product back-office administration pages, user login, and registration pages. Finally, Sect. 5 concludes this paper and discusses future research plans.

2 Structure Design of Fresh Products Application Information Management System

2.1 Requirements Analysis of System Functional

Fresh Products Application Information Management System aims to improve the efficiency of new product information management. The fresh food supermarket app system needs to be divided into the administrator background and the mobile phone client. The supermarket administrator uses the administrator background to manage the supermarket's goods and users' orders; Consumers use the mobile phone client to provide supermarket consumers with online supermarket commodity purchase services.

(1) *Administrator Functional Requirements*

Supermarket login function for administrators. System administrator background supermarket administrator login and logout. The process of commodity management. Commodity addition, commodity maintenance, and commodity discount are among its three features. Due to the wide variety of commodities in the supermarket, while adding items, the price, quantity, and classification of things are also added. This makes it easier to manage and view commodities. Because the supermarket sells various products, each of which has a varying shelf life, it is necessary to determine quickly whether the products are in the critical status and alter their prices accordingly. Therefore, commodity maintenance allows the supermarket administrator to modify the commodity information easily and timely to achieve the consistency of online and offline commodity information—order management function. The administrator can view the user's order information, including the goods purchased by the user, delivery address, and contact phone number, and perform logistics delivery for the goods purchased by the user. For the user management function, the administrator can view the user's personal information in the current system, including the user's name, address, and phone number.

(2) *Customer Functional Requirements*

Account login functionality for customers. Use the app platform to log in and out—a feature for product browsing. When a user accesses the mobile phone client's main module, the home page shows classification details for every item. It explores icons for everything at the bottom of the main interface—a function for classifying commodities. The app's home screen displays a classification of products, including fresh items like fruits, vegetables, and seafood, so users can easily find the details of the products they need to buy—the function of product promotion. On the home page of the mobile phone client, display the real-time product promotion information released by the supermarket administrator so that customers can directly find the purchase path of promotional products. Personal information management function.

When users use this system for shopping, they need to log in to their accounts. For the first time, they can register first. The registration contents include the user name, password of the personal statement, user gender, user address, user telephone, and other information. If accounts forget their passwords when logging in to their accounts, they can retrieve them. After logging in, they can view historical orders and shopping carts.

Shopping cart management function: The shopping cart that users can add after browsing the products. The shopping cart contains the product information that users have added in the past. Accounts can delete and continue to add products, or they can directly click purchase from the shopping cart and enter information to submit an order—commodity purchase function. After the user selects the goods he needs, he can directly purchase them in the shopping cart and submit the order. After the order is submitted successfully, the user must bind a bank card to pay for the order. At this time, the user's bank account will be associated with the user's account.

2.2 System Structural Design

As shown in Fig. 1, the background management module is mainly used by supermarket administrators to manage supermarket commodities, commodity purchases, user orders, and user information. First, the Fresh Food Supermarket App is an online supermarket system. The administrator needs to add the existing inventory and goods sold in the offline supermarket at this time, including the name, selling price, purchase price, picture of the interests, category of the goods, and other information into the background management system. Secondly, the supermarket administrator can modify the data on supermarket goods in different periods. For example,

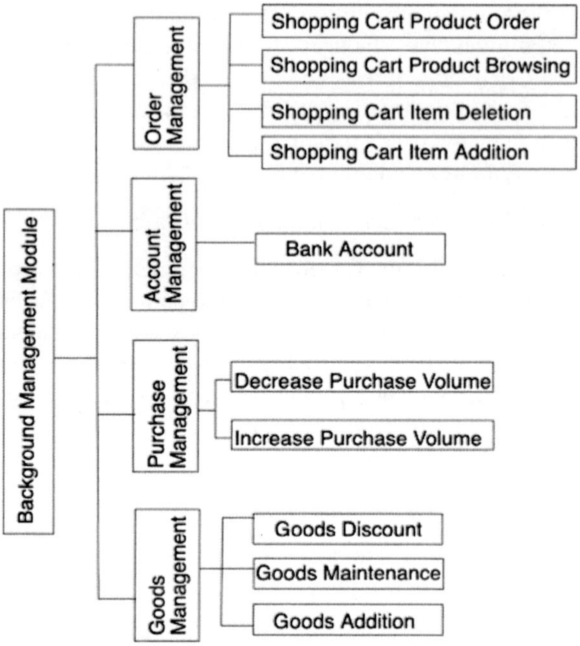

Fig. 1 System function Structural

if the goods are on the way, it is necessary to update the selling price, pictures, and other information about them in real time. Thirdly, the supermarket administrator can understand the market of the goods through the sales of the goods to decide whether to increase or reduce the purchase of the goods in the next purchase. Finally, the supermarket administrator can view the user's information and then view the user's order information. The order information includes the user's shopping cart management. This is mainly for users to directly add the goods to the shopping cart when browsing the goods information on the client after logging in their account, or now enter the shopping cart to browse the added goods information when logging in the next time, then directly place an order in the shopping cart to purchase goods.

3 Database Design of Fresh Products Application Information Management System

3.1 Introduction to Tools Used

In this paper, MySQL-8.0.30 is the software the author uses to create tables, insert data and query the database. The author combines it with the visualization interface of Navicat 16.0.12 to create the database in the visualization interface of Navicat. For the administrator's background commodity management page, user registration page, and user login page, the author uses the software Visual Studio Code. The author uses HTML and CSS to design the interface and add regular expressions to the registration page and login page to determine whether the information filled in by users in the registration or login box meets the required format.

3.2 Database Description

The system mainly includes a user table, commodity table, shopping cart table, order table, and commodity comment table. The data tables are related to each other to complete the user's functions of searching commodities, adding shopping carts, placing orders, and paying. The database has three types of table relationships: one-to-one, one to many, and many to many. After getting familiar with the data, establish the relationship between relational tables, as shown in Fig. 2.

An account can pay for more than one order, and an order can only be paid for by the statement that submitted this order, so the history and the order are one-to-many. One of the accounts can only have a shopping cart, in which the account can add any product he likes. Still, once the product exceeds a specific value, he will be asked to

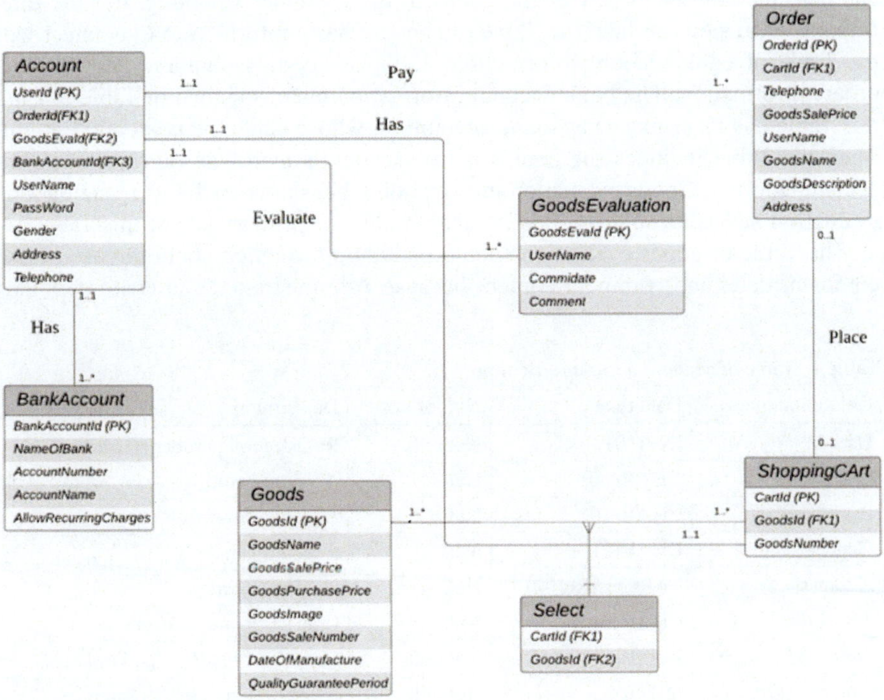

Fig. 2 E-R diagram

delete part of the product and add it again, so the relationship between the account and the shopping cart is one-to-one. An account can comment on all the products he has purchased, and the administrator can trace the analysis back to the account who commented and his order. Hence, the relationship between history and the comment is one-to-many. When a customer wants to pay for an order, the user account needs to be tied to one or more bank cards to pay, so the relationship between the account and the bank account is one-to-many. Multiple shopping carts can select an item with different ids, and a shopping cart can also select multiple objects of various types, so the thing and the coach are in a many-to-many relationship. A submitted coach becomes an order, so the carriage and the order have a one-to-one relationship.

3.3 Logical Structure Design of the Data

The account basic information table is shown in Table 1. Firstly, the user id is the primary key in this Table, which is not allowed to be null, and its data type is int. Secondly, every account needs a username and password to login into the application of the fresh product, and their data type is char. Then the Table should also include the accounts' information about their gender, address, and telephone number, which can help the courier to deliver the goods to the customer smoothly. For the three foreign keys, they can help the Table of account basic information to connect with the Table of bank account information, Table of goods evaluation, and Table of orders information. The bank account information table is shown in Table 2. First, this Table has its primary key–bank account id, whose data type is int and not null. Then the Table includes the bank's name, account's name, account number, and account address. Goods evaluation information table shown in Table 3, each good's evaluation has its id, so the goods evaluation id is the primary key of this Table.

The Table of goods evaluation should include information about the user name, comment date, and comment content because Administrators and customers have

Table 1 Table of account basic information

Column name	Data type	Null or not	Description
User Id	INT(10)	Not null	Registration number (primary key)
User name	CHAR(10)	Null	Account name
Pass word	CHAR(10)	Null	Account Password
Gender	CHAR(2)	Null	Account gender
Address	VARCHAR(100)	Null	Delivery address
Telephone	CHAR(11)	Null	Contact phone number
Order Id	INT(10)	Not null	Order number (foreign key1)
Goods eva Id	INT(10)	Not null	Evaluation number (foreign key2)
Bank account Id	INT(10)	Not null	Bank card number (foreign key3)

Table 2 Table of bank account information

Column name	Data type	Null or not	Description
Bank account Id	INT(10)	Not null	Bank card number (primary key)
Name of bank	CHAR(50)	Null	Bank name
Account name	CHAR(50)	Null	Account Name
Account number	CHAR(10)	Null	Account Number
Address	VARCHAR(100)	Null	Delivery address

Table 3 Table of goods evaluation information

Column name	Data type	Null or not	Description
Goods eva Id	INT(10)	Not null	Evaluation number (primary key)
User name	CHAR(10)	Null	Account name
Comment date	CHAR(10)	Null	Comment date
Comment	VARCHAR(200)	Null	Comment content

the right to know what is said in the comment and when and by whom the statement is sent. Especially for a qualified administrator, he should timely find and delete comments containing prohibited words. As shown in Table 4, the orders information table uses order id as its primary key. When a user selects an item from his shopping cart and submits it, it will be saved as an order for payment. So the Table naturally contains the cart id as its foreign key to help it to connect with the Table of the shopping cart. The information of an order must include the trade names of all the goods in the order, the user name, and the phone number used by the user to place the order so that the administrator can contact the user corresponding to the order on time when the inventory is insufficient, and the order needs to be adjusted. In addition, the order should also display the total amount of the goods in the order and a brief description of the goods in the order so that users can check their orders before placing an order. The shopping cart information table is shown in Table 5, and the card id is the primary key for this Table. To add items to the shopping cart, the Table adds the goods' id as its foreign key to connect with the Table of goods information.

Furthermore, it also adds the int data of the goods number. The goods information table is shown in Table 6, and it has the primary key–goods id, which is the int type of the data. Table of goods information is mainly built for the administrators. It includes the information about the goods sale price, the goods purchase price, and the goods sale number so the administrator can see the profit and sales of goods so that the next purchase can be based on this to increase or reduce the purchase volume of a specific type of goods. Of course, the product information table must contain the product name and a brief description so that administrators and customers can thoroughly understand the product. In the last and most important design, the author adds goods information about the date of manufacture and the quality guarantee period into this

Table 4 Table of orders information

Column name	Data type	Null or not	Description
Order Id	INT(10)	Not null	Order number (primary key)
Telephone	CHAR(11)	Null	Contact phone number
Goods sale price	FLOAT(5,2)	Null	Goods sale price
User name	CHAR(10)	Null	Account name
Goods name	CHAR(50)	Null	Goods name
Goods description	VARCHAR(100)	Null	Product description
Cart Id	INT(10)	Not null	Cart number (foreign key)

Table 5 Table of shopping cart information

Column name	Data type	Null or not	Description
Cart Id	INT(10)	Not null	Cart number (primary key)
Goods number	INT(10)	Null	Goods addition number
Goods Id	INT(10)	Not null	Goods id number (foreign key)

Table so that the administrator can query the database information to know whether the goods are on schedule and promote the products on schedule. This can reduce unnecessary food waste, and the cost of handling expired food for supermarkets.

Table 6 Table of goods information

Column name	Data type	Null or not	Description
Goods Id	INT(10)	Not null	Goods id number (foreign key)
Goods name	CHAR(50)	Null	Goods name
Goods sale price	FLOAT(5,2)	Null	Goods sale price
Goods purchase price	FLOAT(5,2)	Null	Goods purchase price
Goods description	VARCHAR(100)	Null	Product description
Goods sale number	INT(10)	Null	Goods sale number
Date of manufacture	CHAR(10)	Null	Goods manufacture date
Quality guarantee period	CHAR(10)	Null	Quality guarantee period

4 Implementation of Student Information Management System

4.1 Establishment of Database

After getting familiar with the relationship between relational tables, create six tables. The following is some data that the author has inserted into the Table the author has created. This data will be used as an example to demonstrate some of the query functions the author wants to implement in the database.

In Fig. 3, the author inserts the information about the account in the order in which the author created the table attributes. They are '925', 'Lisa', '000000', 'F', 'Room621, Building6, Lieshan District, Huaibei City, Anhui Province, '15555205525', '001', '2', '456' and '926', 'Peter', '111111', 'M', 'Room401, Building11, Bengshan District, Bengbu City, Anhui Province', '15,655205525', '002', '1', '123'. For Fig. 4, in the same sequence, as the author created the characteristics for the Table of bank account information, the author inserted the bank account information, which is '123', 'China Construction Bank,' 'Wu Wang,' '001', 'Room401, Building11, Bengshan District, Bengbu City, Anhui Province' and '456', 'China Construction Bank,' 'San Zhang,' '002', 'Room621, Building6, Lieshan District, Huaibei City, Anhui Province'. In Fig. 5, the author has inserted a series of attribute values in the order of the Table's attributes, which are '1', 'Peter,' '2022-9-24', 'This kind of snack is very delicious!' and '2', 'Lisa,' '2022-9-25', 'This tastes terrible!'. the author has added a number of attribute values to Fig. 6 in the sequence of the table's attributes, which are '001', '15555205525', '19.89', 'Lisa', 'Shine Muscat', 'Grapes from Yunnan province', '001' and '002', '15655205525', '164', 'Peter', 'Boston Lobster', 'Lobsters from Canada', '002'. In Fig. 7, the author has inserted a series of attribute values in the order of the Table's attributes. They are '001', 2, '111', '002', 1, '112' and '003', 3, '111'. In Fig. 8, the author inserts the information about the account in the order in which the author created the table attributes. They are '111', 'Shine Muscat', '8.98', '6.98', 'Grapes from Yunnan province', '3000', '2022-9-23', '5 days' and '112', 'Boston Lobster', '164', '120', 'Lobsters from Canada', '500', '2022-9-24', '3 days'.

UserId	UserName	PassWord	Gender	Address	Telephone	OrderId	GoodsEvald	BankAccountId
925	Lisa	000000	F	Room621,Building6,LieshanDistrict,HuaibeiCity,AnhuiProvince	15555205525	1	2	456
926	Peter	111111	M	Room401,Building11,BengshanDistrict,BengbuCity,AnhuiProvince	15655205525	2	1	123

Fig. 3 Account table

BankAccountId	NameOfBank	AccountName	AccountNumber	Address
123	China Construction Bank	Wu Wang	001	Room401,Building11,BengshanDistrict,BengbuCity,AnhuiProvince
456	China Construction Bank	San Zhang	002	Room621,Building6,LieshanDistrict,HuaibeiCity,AnhuiProvince

Fig. 4 Bank account table

GoodsEvaId	UserName	CommentDate	Comment
1	Peter	2022-9-24	This kind of snack is very dilicious!
2	Lisa	2022-9-25	This tastes terrible!

Fig. 5 Goods evaluation table

OrderId	Telephone	GoodsSalePrice	UserName	GoodsName	GoodsDescription	CartId
1	15555205525	19.89	Lisa	Shine Muscat	Grapes from Yunnan province	1
2	15655205525	164.00	Peter	Boston Lobster	Lobsters from Canada	2

Fig. 6 Table of orders table

CartId	GoodsNumber	GoodsId
1	2	111
2	1	112
3	3	111

Fig. 7 Table of shopping cart table

4.2 *Query Function Implementation*

Administrators can query the database to get the order number of the poor evaluation and then get the specific information of the order according to the order number. For example, as shown in Fig. 9, administrators can look up the goods evaluation id of the comment with the keyword terrible and finds the order id associated with the comment. In this way, administrators can timely make necessary adjustments to the comments with prohibited words. Furthermore, as shown in Fig. 10, administrators

GoodsId	GoodsName	GoodsSalePrice	GoodsPurchasePrice	GoodsDescription	GoodsSaleNumber	DateOfManufacture	QualityGuaranteePeriod
111	Shine Muscat	8.98	6.98	Grapes from Yunnan province	3000	2022-9-23	5 days
112	Boston Lobster	164.00	120.00	Lobsters from Canada	500	2022-9-24	3 days

Fig. 8 Goods table

OrderId
1

Fig. 9 The query of Special Comment

OrderId	Telephone	GoodsSalePrice	UserName	GoodsName	GoodsDescription	CartId
1	15555205525	19.89	Lisa	Shine Muscat	Grapes from Yunnan province	1

Fig. 10 The query of Order Information

can also find out the order's details by the order ID and contact the comment to make an appropriate reply and deal with it.

SELECT OrderId FROM Account WHERE GoodsEvaId IN (SELECT GoodsEvaId FROM GoodsEvaluation WHERE Comment LIKE '%terrible!');

SELECT * FROM Orders WHERE OrderId = '1';

Administrators should always pay attention to goods with a short shelf life, especially some fresh products. As shown in Fig. 11, administrators can query the information about the quality guarantee period of the products. Administrators must promote items to be sold out as quickly as possible, as shown in Fig. 12, approaching the shelf life to reduce unnecessary waste.

SELECT GoodsId, GoodsName FROM Goods WHERE QualityGuaranteePeriod <= 3;

UPDATE Goods SET GoodsSalePrice = 150 WHERE GoodsId = '112';

SELECT * FROM Goods WHERE GoodsId = '112'.

Administrators can classify the orders according to the same province so that administrators can directly retrieve the goods from the warehouse near the area, effectively avoiding unnecessary waste during transportation. As shown in the example in

GoodsId	GoodsName
112	Boston Lobster

Fig. 11 The query of Expiry Products

GoodsId	GoodsName	GoodsSalePrice	GoodsPurchasePrice	GoodsDescription	GoodsSaleNumber	DateOfManufacture	QualityGuaranteePeriod
112	Boston Lobster	150.00	120.00	Lobsters from Canada	500	2022-9-24	3 days

Fig. 12 Updated Product Price Information

OrderId	Telephone	GoodsSalePrice	UserName	GoodsName	GoodsDescription	CartId
1	15555205525	19.89	Lisa	Shine Muscat	Grapes from Yunnan province	1
2	15655205525	164.00	Peter	Boston Lobster	Lobsters from Canada	2

Fig. 13 Search for information on products in the same area

GoodsId	GoodsName	GoodsSaleNumber
111	Shine Muscat	3000

Fig. 14 Check the most popular items

Fig. 13, the administrator has grouped orders belonging to the same province using the province keyword.

SELECT * FROM Orders WHERE OrderId IN (SELECT OrderId FROM Account.

WHERE Address LIKE '%AnhuiProvince');

As shown in Fig. 14, administrators can query the database for the goods name, goods id, and specific sales number of the goods with the most significant sales in the goods table. This way, administrators can selectively add or delete the purchased quantity of a particular commodity during the next warehouse purchase.

SELECT GoodsId,GoodsName,GoodsSaleNumber FROM Goods.

WHERE GoodsSaleNumber = (SELECT MAX(GoodsSaleNumber) FROM Goods).

4.3 *Implementation of Main Modules*

As shown in Fig. 15, in "Goods Management," app the administrators can click "Goods Addition" to enter commodity information under the stock of the current supermarket. They can classify the commodities, such as fruits, vegetables, seafood, and quick ice products, and specify if the item is now promotional in addition to providing the product's name and price.

As shown in Fig. 16, the information entered by the user during account registration will be saved directly in the account table in the database. Therefore, after the customer completes the order, administrators can find the user's phone number and address information through the backend query so that when the goods are delivered, it will be convenient for the courier to be able to contact the customer in time to prevent damage to the fresh produce due to the courier's failure to contact the customer in time for the problem.

As shown in Fig. 17, users can log in through the login page after registering information. Because regular expressions are added to the page, if the account and password entered do not conform to the set rules, the text box will flash red to prompt users to pay attention.

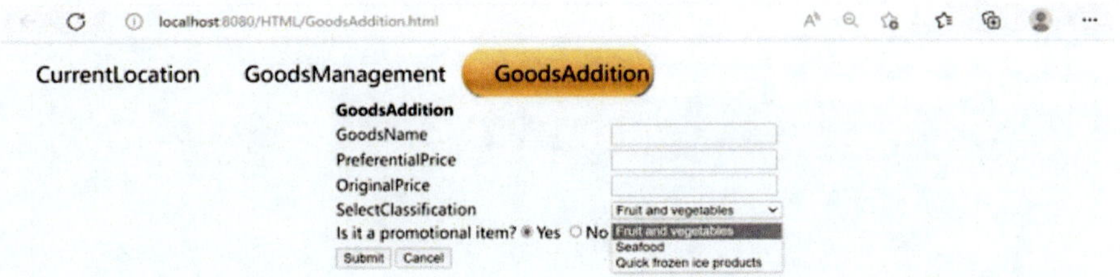

Fig. 15 Goods add page

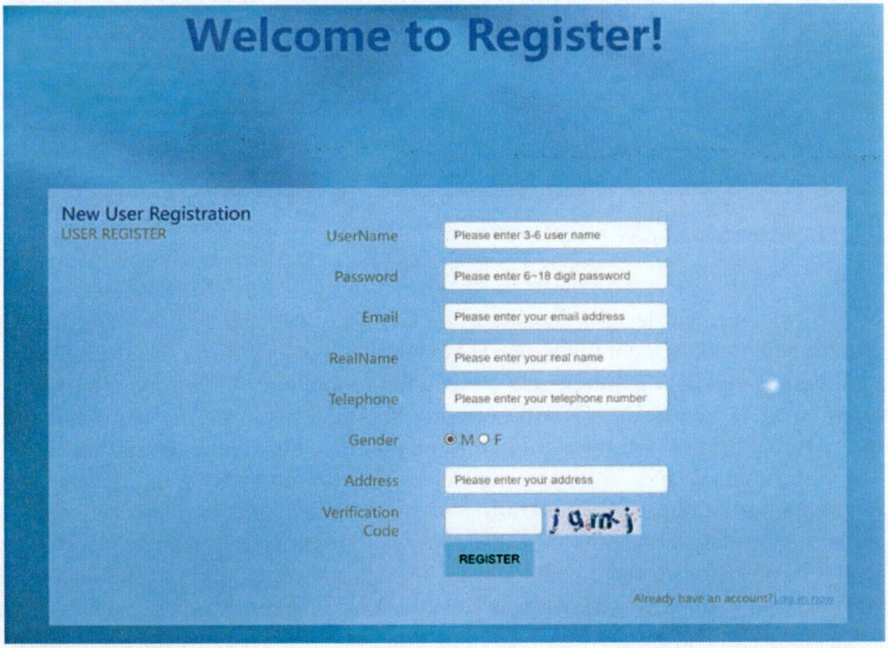

Fig. 16 Accounts register page

5 Conclusions

After understanding some online supermarket software, the author finds that most supermarket software does not manage fresh produce reasonably well and often causes a lot of unnecessary food waste due to product management problems. Therefore, this paper designs and implements a new product application information management system to help manage fresh produce and reduce resource waste due to improper management. The author first analyzes the functional requirements

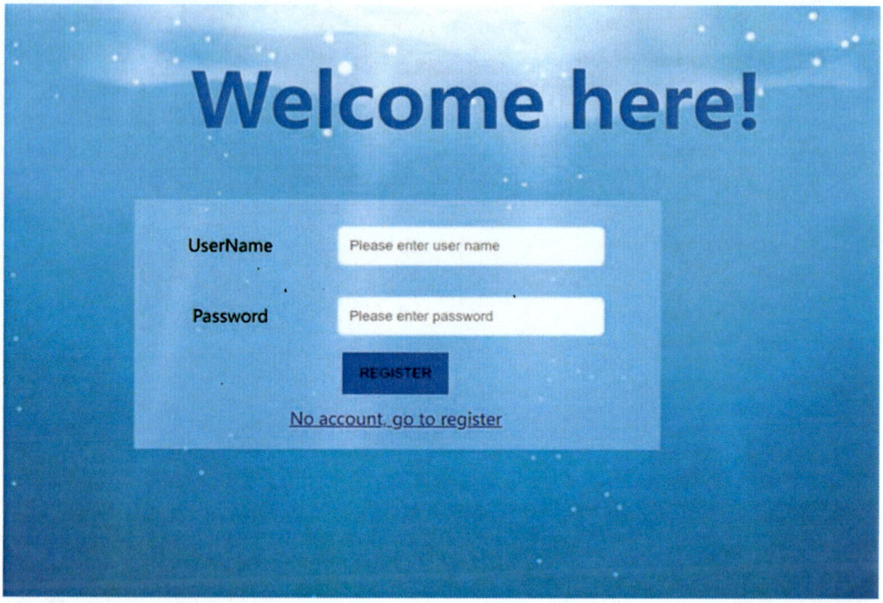

Fig. 17 Accounts login page

of the system, then designs the structural framework of the system based on the requirements analysis, completes the ER diagram based on the users' and administrators' understood data requirements, processing requirements, security and integrity requirements, creates the database, creates the tables, inserts the data, and then uses MySQL to perform the desired query functions. Finally, by compiling HTML and CSS in the VScode software, the web design for the background management of product additions, user registration accounts, and login accounts was accomplished. The main contribution of this paper is to improve the fresh produce application information management system. The administrator through the query can be timely according to the customer's evaluation to make appropriate adjustments to commodity management. And when dealing with the issue of expired goods, the administrator can also make timely price reduction processing, which is not only conducive to reducing unnecessary food waste, but also allows customers to buy food at a discounted price. The current study still has certain shortcomings. Some of the inquiry functions require a certain level of operational skills on the part of the administrator, and the handling of expired fresh produce is only considered at the stage of selling the product. Future research should consider avoiding unnecessary fresh produce waste while transporting products and simplify the operation of database queries as appropriate.

References

Aydin A et al (2017) Save the Kiwi: encouraging better food management through behavior change and persuasive design theories in a mobile app. In: Proceedings of the 2017 CHI conference extended abstracts on human factors in computing systems

Baarsma B, Groenewegen J (2021) COVID-19 and the demand for online grocery shopping: empirical evidence from the Netherlands. De Economist 169(4):407–421

Chao T (2020) Study on the optimization of the business model of supermarket fresh O2O platform

Eliaz K, Weisburd S, Oren-Kolbinger O (2021) How do shoppers respond to information on price changes in multiple product categories? Evidence from a field experiment in online supermarket shopping

Freeman M, Hyland P, Soar J (2003) Australian online supermarket usability. In: Proceedings of the 4th annual conference of the ACM special interest group on computer-human interaction

Hanus G (2016) Consumer behavior during online grocery shopping. In: CBU international conference proceedings, vol 4

Mkoba ES (2023) Development of supermarket management information system

Taban J, Bakulembeze T (2013) An online mobile shopping application for Uchumi supermarket in Uganda

Xiang X et al (2014) Fresh food online supermarket development study. J Electron Commer Org (JECO) 12(2):14–30

Xu L, Zhu C (2009) Research and development on the chain supermarket management system based on B/S. In: 2009 2nd international conference on power electronics and intelligent transportation system (PEITS), vol 3